EMS Documentation

John Snyder, RN, EMT-P

Please visit Brady's Companion Website for this text at
www.prenhall.com/snyder.

PEARSON
Prentice
Hall

Upper Saddle River, New Jersey 07458

Library of Congress Control Number: 2007940946

Publisher: Julie Levin Alexander
Publisher's Assistant: Regina Bruno
Executive Editor: Marlene McHugh Pratt
Senior Managing Editor for Development: Lois Berlowitz
Development Editor: Triple SSS Press Media Development, Inc.
Director of Marketing: Karen Allman
Executive Marketing Manager: Katrin Beacom
Marketing Specialist: Michael Sirinides
Managing Editor for Production: Patrick Walsh
Production Liaison: Faye Gemmellaro
Production Editor: Karen Fortgang, bookworks publishing services

Media Product Manager: John Jordan
Media Project Manager: Stephen Hartner
Manufacturing Manager: Ilene Sanford
Manufacturing Buyer: Pat Brown
Senior Design Coordinator: Christopher Weigand
Interior Design: Amy Rosen
Cover Design: Christopher Weigand
Cover Photography: EMSA, Tulsa, Oklahoma
Composition: Aptara, Inc.
Printing and Binding: Bind-Rite Graphics
Cover Printer: Phoenix Color Corporation

NOTICE

The author and the publisher of this book disclaim any liability, loss, or risk resulting directly or indirectly from the suggested procedures and theory, from any undetected errors, or from the reader's misunderstanding of this reference guide. It is the reader's responsibility to stay informed of any new changes or recommendations made by any federal, state, and local agency/organization as well as by his/her employer or medical director.

EMS Documentation provides EMS personnel with principles to guide in PCR documentation. It is not intended to replace local practice standards or protocols. All EMS personnel are advised to follow the guidelines set forth by their medical director and administrators.

The examples provided throughout the text are only intended to illustrate the content presented and are not to be used as templates for documentation. *EMS Documentation* is not intended in any manner to provide direction or advice in matters relating to the billing of EMS services and reimbursement. All EMS organizations are encouraged to have active compliance programs and to seek the counsel of a qualified EMS attorney for matters relating to PCR documentation and reimbursement.

Pearson Prentice Hall™ is a trademark of Pearson Education, Inc.
Pearson® is a registered trademark of Pearson plc
Prentice Hall® is a registered trademark of Pearson Education, Inc.

Pearson Education Ltd., London
Pearson Education Singapore, Pte. Ltd.
Pearson Education Canada, Inc.
Pearson Education—Japan
Pearson Education Australia, Pty. Ltd.

Pearson Education North Asia, Ltd., Hong Kong
Pearson Educación de Mexico, S.A. de C.V.
Pearson Education Malaysia, Pte. Ltd.
Pearson Education, Upper Saddle River, New Jersey

10 9 8 7 6 5 4 3 2 1
ISBN 13: 978-0-13-236964-0
ISBN 10:   0-13-236964-8

This book is dedicated to my sons, Zachary Snyder and Micah Snyder. Thank you for giving so this book could be written.

Contents

Chapter **3**

Documentation and the Financing of EMS   31

Chapter **4**

Legal Responsibilities and EMS Documentation   45

Chapter 5

Documentation Fundamentals 63

Chapter 6

Essential Documentation Elements 77

Chapter 7

Narrative Documentation 105

Chapter 8

Medical Necessity 119

Chapter 9

Putting It All Together   141

Chapter 10

Patient Refusals   149

Chapter **11**

Incident Reporting   163

Chapter **12**

Verbal Reports   173

Chapter 13

Complex Patient Encounters   185

Appendix A

Patient Care Report (PCR)   195

Appendix B

Standard Charting Abbreviations   209

Answers to Case Study Questions   213

Answers to Chapter Review Questions   217

Answers to Critical Thinking Discussion Exercises   237

Glossary   243

Index   245

Preface

Many in EMS, at all levels of provider training, have been well trained in patient assessment and management skills but not as well trained in the skill of documentation. A review of most EMS textbooks will reveal a section, perhaps even a chapter, dedicated to documentation. While working diligently to perfect and maintain clinical skills, EMS professionals seem to struggle with documentation. The skill that supports everything that is taught in the text is reduced to a section or chapter, and this is one reason why so many in EMS struggle with documentation.

Attaining proficiency in documentation is vital to EMS. The EMS industry today is facing the challenges of reimbursement, quality management, and legal and governmental accountability. It's easy for EMS administrators to seek a silver bullet to answer these challenges when a common denominator is right under the noses of their staff each and every day—the Patient Care Report.

Attaining proficiency in documentation is vitally important to protecting and advancing the careers of EMS professionals. As a professional you are evaluated through the window of documentation, and your patient care is viewed through the window of your documentation. When all is said and done, the reality is this: if great patient care is not documented, then great patient care was never provided.

The objective of *EMS Documentation* is to assist the EMS professional, whether novice or expert, in attaining proficiency in documentation. Whether you are new to EMS or have years of experience, this text is for you.

The chapter features include:

- **Key Ideas:** Outline the objectives of the chapter.
- **Case Studies:** Apply EMS case studies to developing documentation skills. This feature will highlight the key concepts throughout the text, portray common documentation errors, and demonstrate appropriate application of each chapter's core material to EMS practice.
- **Case Study Questions:** At the beginning of each chapter involve the reader with critical thinking concepts that may be encountered in the real world of EMS calls.
- **Introduction:** An overview provided to help the reader gain insight into core concepts presented in each chapter.
- **Key Terms:** Terms highlighted in bold throughout the chapter and defined in the margins for better student comprehension.
- **On Target:** Pearls of wisdom that reinforce chapter content.
- **Chapter Review Questions:** Reinforce and quiz the reader on what was learned in the chapter.
- **Critical Thinking Discussion Exercises:** Critical thinking is vital to the delivery of excellent patient care. This feature applies critical thinking to Patient Care Report (PCR) documentation to assist the reader in proper documentation of the growing complexities of patient care.
- **Action Plan:** Promotes application of the core material presented in each chapter.
- **Practice Exercises:** Provide the reader the opportunity to practice the content presented in each chapter.

EMS Documentation is a worktext. There are plenty of exercises so you can practice the material presented in each chapter and, more importantly, to make you think about documentation. You really don't have to struggle with documentation. You can become proficient, and in doing so you will protect and advance your career and the EMS profession.

John Snyder, RN, EMT-P

Acknowledgments

This text would not have been possible without those who invested in the project and the author. I must extend thanks to:

- Marlene Pratt and the staff at Pearson Education/Brady for their investment in the development and advancement of EMS.
- Alice "Twink" Dalton and Jeanne O'Brien for imprinting into their students a pattern of excellence and professionalism on which to build EMS careers.
- Andrea Edwards for her support and direction through the process of completing this text. You have been the best "coach."
- The truly remarkable EMS professionals I had the privilege of working with over the years at Acadian Ambulance, EMSA, and Muskogee County EMS.
- Don Elkins, Bill Humphrey, and Terri Mortensen for the leadership, professionalism, and integrity they modeled as my superiors.
- Keith Simon of Acadian Ambulance for his support of this project.

Most importantly, I thank my wife Elaine for her encouragement, patience, and unwavering support.

Reviewers

Special thanks to the following reviewers. I greatly appreciate the manner in which you applied your own experience and expertise to review of this text. Your observations, comments, and suggestions redirected and refined the development of *EMS Documentation*.

John Beckman, AA
EMS Instructor
Technical Center of DuPage County
Addison, IL

Rodney K. Cantrell, CCEMT-P
Coordinator/Instructor
Morgan Emergency Training
West Liberty, KY

Deb Cason
UT Southwestern Medical Center
Emergency Medicine Education
Dallas, TX

Heather Collins, EMT-I
EMT Program Director/AHA Faculty
Treasure Valley Community College
Ontario, OR

Steven Creech, EMT-P
EMS Instructor/Training Officer
Washington County EMS
Plymouth, NC

Tim Duncan, RN, CCRN, CEN, CFRN, EMT-P
Flight Nurse/Educator
St. Vincent Life Flight
Toledo, OH

Sandra Hartley, MS, CP
Paramedic Program Director
Pensacola Junior College
Pensacola, FL

Scott C. Jones, MBA, EMT-P
Director, Paramedic Academy
Chairperson, Allied Health
Victor Valley College
Victorville, CA

Bill Locke
EMS Instructor/Coordinator
Moraine Park Technical College
Fond Du Lac, WI

Brittany Martinelli, MHSc, BSRT, NREMT-P
Assistant Professor EMS
Santa Fe Community College
Gainesville, FL

Ronald Mrochko, NREMT-P
Adjunct Instructor
Trans-Med Ambulance, Inc.
Scranton, PA

John Rinard, BBA
Instructor
TEEX EMS Program
College Station, TX

Sandy Waggoner, EMT-P
Public Safety Coordinator
EHOVE Joint Vocational School District
Milan, OH

John Snyder

About the Author

John Snyder is an EMS consultant specializing in education, program, and staff development and reimbursement matters. A graduate of Creighton University's Paramedic Training Program in Omaha, Nebraska, and Rogers State University in Claremore, Oklahoma, John has been in EMS for 25 years.

John has a strong clinical foundation, with over 22 years of EMS practice, and has served as a paramedic program instructor, field and communications supervisor, education manager, director of operations, critical care RN, and as a member of the medical review management team of one of the nation's largest Medicare Part B carriers.

CHAPTER 1

Documentation:
What's in It for You?

Key Ideas

Upon completion of this chapter, you should know that:

- As a relatively new profession, Emergency Medical Services has made remarkable advancements. Despite these advancements, significant challenges face the EMS profession today that have a direct link to PCR documentation.

- Documentation is a core EMS skill. Becoming proficient in documentation will have a positive impact on your career.

- There is a lot at stake each time a Patient Care Report (PCR) is completed.

- Improving the quality of EMS documentation will assist the profession in meeting the challenges facing EMS.

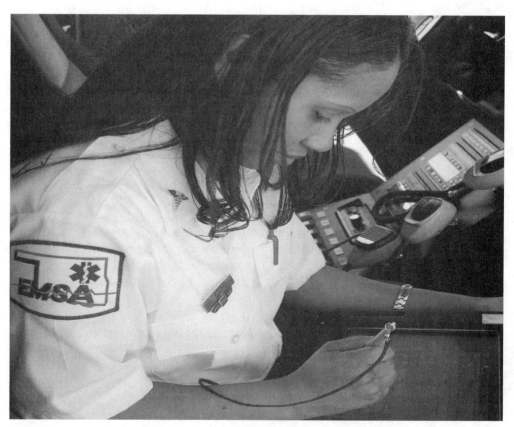

FIGURE 1-1
Paramedic completing PCR (Courtesy EMSA, Tulsa, OK)

CASE
Study

Y ou are dispatched to the scene of a motor vehicle collision (MVC). The location, a rural tree-lined county highway with numerous sharp curves, is well known to you. Your service is called so frequently to this location that it's known as the "trauma referral center." Arriving "on scene," you begin your initial scene evaluation, observing that a late-model pickup has struck a tree at an apparently high rate of speed. As you exit the ambulance, Emergency Medical Responders confirm that the patient whom you note to be slumped over the steering wheel is the only patient at this scene.

You meticulously assure scene safety, evaluate mechanism of injury, and begin your initial assessment, which reveals an unresponsive adult male patient. As the Emergency Medical Responders prepare extrication and immobilization equipment, you complete a thorough rapid trauma assessment, confirming a critically injured multisystem trauma patient. You immediately intervene with appropriate initial airway management, extrication, and textbook perfect head-to-toe spinal immobilization. Within minutes you are in the ambulance and are on the way to the nearest Level One Trauma Center.

En route to the hospital you provide definitive airway management via endotracheal intubation, confirm tube placement, initiate end-tidal CO_2 monitoring, establish bilateral large bore IVs, and provide appropriate fluid volume resuscitation. During transport, you continually monitor and reassess the patient. Each time the patient is moved, you appropriately verify correct placement of the endotracheal tube. Ten minutes from the receiving hospital you provide the receiving facility a concise radio report to begin the transfer of care process.

Arriving at the receiving facility, you provide the receiving RN an appropriate transfer of care report. However, despite perceived textbook-perfect care during the 20-minute transport, the patient rapidly deteriorates into cardiac arrest. Failing to respond to aggressive treatment per your state's EMS protocols, your patient is pronounced dead by the Emergency Department physician 15 minutes after your arrival at the Emergency Department. As you leave "Trauma Room One," you note that the staff is handling the patient quite roughly as they struggle to remove the remainder of his clothing.

This was a challenging call, but you were in "the zone" the entire time. Despite the patient's poor outcome, you performed every skill very well. You consider yourself good at what you do, and you have earned the respect of your peers. Quite tired, you go out to the ambulance to help your partner get things ready for the next call. It was an unfortunate outcome, but you know you, your partner, and the other EMS personnel did everything that could have been done for this patient. You spent many hours refining the skills that you used to treat this patient, and they couldn't have been performed any better. The call is over, or is it?

Questions

We begin in EMS with assessment. Therefore, it is fitting that we begin the discussion of EMS documentation by assessing your thoughts and views on this subject. Please refer to Answers to Case Study Questions at the back of this book.
1. What is your definition of EMS documentation?

2. How does documentation match up to other EMS skills such as spinal immobilization and airway management?

3. What is the purpose of the Patient Care Report?

Patient Care Report (PCR)

The Patient Care Report (PCR) is the professional documentation tool of the EMS professional. Previously known as the "trip sheet," the PCR records demographic, financial, and patient care information. Also referred to as the "Prehospital Care Report."

ON TARGET EMS documentation is the record of the unique professional activities transforming clinical judgments and interventions into a professional, legal, and financial document.

Introduction

Documentation is most often synonymous with the "trip sheet" or the **Patient Care Report (PCR)**, although it actually means different things to different people. For many, preparing the Patient Care Report is something that is done, but not understood. To some, completing this report is nothing more than the last tedious task at the end of a call.

Documentation has often been extremely confusing for EMS. Many Paramedics, Advanced EMTs, EMTs, and Emergency Medical Responders[1] have differing perspectives on what EMS documentation is and how it should be done. Direction from medical directors, quality managers, and administrators often conflict, and clinical and financial interests can seem to be in opposition. From state to state, county to county, and even service to service, the format of the PCR or trip sheet differs. Adding to the confusion is the ever increasing variety of electronic PCRs in use in many EMS systems.

Precisely stated, EMS documentation is the record of the unique professional activities transforming clinical judgments and interventions into a professional, legal, and financial document. With this as our working definition, EMS documentation becomes more than a paper form that is filled out, an electronic exercise, a task that completes the call, or a written summary of everything done for the patient. Documentation records the unique contributions EMS professionals make to health care.

This text will take you on a journey beyond the traditional view of EMS documentation. This journey will take you from viewing documentation as simply a closing thought to the realization that documentation is as important as the call itself. If you are new to EMS, this text will assist you in developing documentation skills so you get a good start as you develop in your EMS career. In doing so, you will gain an edge that many before you have not had.

If you are an established EMS provider, this text is designed to extricate you from your documentation comfort zone by providing a new perspective on EMS documentation. With many opinions already engrained as to the purpose and function of documentation in EMS, some of the content may be controversial. Take the journey and allow your views of EMS documentation to be challenged and refined, and in doing so you will experience professional growth that will have significant impact on your EMS career.

KEY TERMS

Note: Page numbers indicate where the following key terms and definitions first appear.

Patient Care Report (PCR)
(p. 3)

Health Insurance Portability
and Accountability Act of
1997 (HIPAA) (p. 5)

Emergency Medical Treatment
and Active Labor Act of
1986 (EMTALA) (p. 5)

reimbursement (p. 5)

Diagnosis Related Groups
(DRGs) (p. 5)

Balanced Budget Act 1997
(BBA) (p. 5)

The Evolution of EMS

Many things have evolved over the past 25 years in EMS. First, the scope of EMS practice has changed. For example:

> *Unit 21, this is Pointe Coupee General Hospital. Go ahead and start an IV of D5W. Run the IV at TKO and continue to transport. We have no further orders at this time.*

---

[1]These are the new designations for the four nationally recognized levels of care, as defined by NHTSA. They replace the designations EMT-Paramedic, EMT-Intermediate, EMT-Basic, and First Responder.

Twenty years ago EMS personnel were the "eyes and ears of the physician," but unfortunately they were just that—eyes and ears. The skills the first EMTs and paramedics were taught were to be used only if a physician had ordered them. All treatment was based on what the physician's "eyes" had seen. Fortunately, this has changed. Today's EMS providers are required to think critically as they assess, evaluate, and make treatment decisions based upon what *they* have seen. This has been an evolutionary process. Today, one wouldn't even think of transmitting an ECG to verify ventricular tachycardia, as was done 20 years ago. Doing so would certainly guarantee a "sit down" with the medical director.

Second, equipment has changed as research has proven that lives are saved through advanced life support interventions initiated prior to arriving at a hospital. Equipment has evolved from a few air splints, a plywood spine board, and a bag of cervical collars to advanced life support equipment that was reserved for the ICU just 20 short years ago.

Third, public awareness of the capabilities of EMS has evolved as the public has been educated on Emergency Medical Services. Today's sophisticated health care consumer now realizes that EMS personnel are more than just "ambulance drivers." A number of years ago, a paramedic and an EMT were at their station when a local physician showed up at the door with the announcement that he had a "sick baby" at his office that needed to be transferred to the PICU. He wanted assurances that one of them would be riding in the back with the baby on the way to the hospital. Several problems were noted rather quickly. First, if he had a sick baby at his office, what was he doing at the EMS station? Second, did he think that the crew was simply going to strap the baby to the stretcher and then leave the infant to fend for himself as they drove to the hospital? The EMT politely assured the physician that one of them would ride in the back at all times. The reality was apparent: even though EMS had changed, the physician's perspective of the capabilities of EMS had not changed over the many years of his practice. When he had begun his practice, having an "attendant" in the back of the ambulance was a valid concern because not all funeral home ambulance attendants did so. Fortunately, public awareness of EMS has changed.

Every time you respond to a call and provide excellence in patient care and customer service, you promote the growth of the EMS profession.

Today's EMS is much different. As a result of the remarkable advancements of the last 40 years, the public has gone from ignorance to expectancy. Today, across the country, the public has the expectation that professional emergency care will be available whenever it is needed. However, along with the advancements brought by the evolution of EMS are significant challenges that face the EMS industry today.

The Challenges Facing EMS

While EMS has experienced tremendous advancement, the foremost challenge of establishing EMS as a profession remains. What is the EMS profession? Is EMS "pre-hospital" or "out-of-hospital" medicine? Is EMS public safety? How does EMS relate to medicine and nursing? Are salaries in EMS an appropriate reflection of the level of responsibility, education, and expertise of today's EMS professional? Ask yourself:

- Are you satisfied with the pay that you receive in EMS? Do you think that EMS pay is consistent with the pay of other health care professions such as nursing? Why?
- Have you ever been called an "ambulance driver"? If this has happened to you, does it bother you? If it hasn't, would it bother you? Why?
- Does the health care community have an accurate understanding of your capabilities as an out-of-hospital care provider? Why?

The everyday experience of most in EMS will reveal that EMS has yet to be established as a health care profession.

The second challenge facing EMS is the increasing legislative, regulatory, and legal involvement in Emergency Medical Services. As EMS has grown, so have the interest and involvement of the legislative and legal communities. Forty years ago there were essentially no regulatory requirements on EMS. There were no laws, regulations, or state EMS offices. During the early years of EMS, a lawsuit against an EMT or a paramedic was unheard of. As EMS has become recognized as an integral part of the nation's health care system, it is impacted by the same health care legislation that hospitals, physicians, and other members of the health care community must deal with. Examples of recent legislation affecting EMS are **Health Insurance Portability and Accountability Act of 1996 (HIPAA)** and **Emergency Medical Treatment and Active Labor Act of 1986 (EMTALA)**. HIPAA and EMTALA have placed greater demands on EMS documentation.

In the days of EMS infancy, life was sheltered. The public was unaware of the development of EMS systems, and Good Samaritan Laws provided an umbrella of protection from the legal accountability reserved for other health care professionals. As public awareness has grown, so has the public's expectation regarding pre-hospital health care, and EMS providers are now attractive legal targets for litigation. As a health care professional you are a potential legal target because the public is watching. It can be a thrill to be at the "big call" and have bystanders stand in awe of your every move. However, with increased public awareness and the expectancy of good outcomes all the time, every call is open to legal scrutiny, which brings us back to documentation.

The Patient Care Report (PCR) is the primary legal tool used to evaluate care versus legal standards. When expectations are not met, it will be asked, "Was the patient care consistent with the care that would have been given by another pre-hospital care provider?" and "Does the patient care line up with standards of care and protocols?" These are questions that will be asked of the PCR documentation, not of the EMS provider.

The third challenge facing EMS is that of **reimbursement**, the manner in which EMS systems and providers are paid. Prior to the 1980s, health care providers essentially had a "blank check" from Medicare and insurance companies. Health care and ambulance providers were paid for what was billed. Providers billed Medicare and insurance companies, and the money rolled in. The implementation of **Diagnosis Related Groups (DRGs)** in the 1980s began the process of setting limits for what would be paid for Emergency Medical Services. As Medicare and insurance companies began using DRGs to decide the amounts that would be paid for EMS services, budgets tightened. As public and private EMS budgets became strained in the 1980s and 1990s, EMS systems became more dependent upon reimbursement from Medicare and third-party payers.

The **Balanced Budget Act of 1997 (BBA)** was a turning point for EMS reimbursement, impacting EMS systems in an unprecedented manner. The Balanced Budget Act mandated a "fee schedule" that dictates payment for EMS services, often at a level below operating costs. Think of the effects of DRGs and the BBA as if your own finances were involved. First, the DRGs place a cap on your salary. Then the Balanced Budget Act comes along and sets your pay at a level below the sum total of your monthly bills, quickly unbalancing your budget. Instead of the blank check of the 1970s and 1980s, the amount of the check for each EMS call is already filled in by Medicare and the insurance companies before the ambulance even leaves the station.

The challenges facing EMS today are formidable. Every EMS provider makes a powerful impact on these challenges by attaining proficiency in EMS documentation. Each time a PCR is completed a deposit or a withdrawal is made in overcoming these challenges. Documentation proficiency displays the attributes of the EMS professional, reduces the threat of litigation, and represents the primary tool for assuring appropriate reimbursement for EMS services. Whether the challenge is establishing the professional identity of the EMS profession, the increasing legal interest in EMS, or the financing of EMS, documentation is the common denominator. Documentation is central to meeting the challenges facing EMS, but what's really in it for you?

EMS Documentation

Attaining excellence is the common goal of EMS professionals. It is very important to understand that attaining proficiency in documentation can have a positive effect on your career. Becoming proficient in documentation is often left out of the equation of excellence.

Table 1-1 The PCR Audience

Hospital	Employer			Reimbursement	
Emergency Department physician	Supervisor	DOT researcher		Claims processor	Plaintiff's attorney
Emergency Department RN	Quality improvement (QI) manager	NTSB researcher		Medical review nurse	Defense attorney
Attending physician	Operations manager	NHTSA researcher		QI review (CERT)	Legal clerks
Staff nurses	Medical director	State EMS staff		Postpay medical review nurse	Juries
Case manager	Billing department	State medical director		Office of Inspector General	Judges
Social worker	Compliance officer	State data collection		Private insurance companies	Appeals judges

You Practice Before a Large Audience

EMS documentation is not a private matter. It is very likely that many people will review your PCR documentation long after the call is finished. When the eye of the public is fixed upon the EMS provider's skill in managing the motor vehicle collision (MVC) on a busy highway or the downed football player at the high school stadium, it is truly exciting. It is another thing altogether when documentation turns into a "public" event. The Patient Care Report has an audience that goes well beyond your partner, supervisors, EMS personnel, and bystanders. Long after the ambulance is back to the station, the EMS encounter continues as documentation is viewed by the PCR audience. You might be surprised at who is reviewing your documentation (see Table 1-1).

Is it really possible for your documentation to be this extensively reviewed? Yes, it is. The PCR is used by physicians and nurses who will be continuing the care initiated in the field. The PCR will be used by your employer, state EMS regulators, and researchers for a variety of compliance and quality purposes. Your PCR might be used to make a reimbursement decision, and review of PCRs by plaintiff's and defense attorneys is more commonplace than one might expect.

The Mirror of Your Professionalism

A simple fact basic to EMS documentation is that your quality improvement officer and medical director are not on the call with you. The physicians, nurses, attorneys, Medicare representatives, and state EMS officials are not among the bystanders at the scene. They don't see the patients you take care of, and they don't see the results of your care and skill. They see you only as reflected in PCR documentation. After the care and transport are complete, all that permanently remains is the written record that you create—the PCR. This is the one opportunity you have as to how they will view you as a professional.

Your Most Important EMS Skill

Documentation is as much a skill as performing an assessment, starting an IV, or managing an airway, and it is developed and perfected in the same manner that proficiency is attained in other skills—through commitment, dedication, practice, and tenacity. Your skills are part of what defines you as a professional, and becoming proficient is engrained into us as we are socialized into the EMS profession. It's a "peer thing." How many times have you heard the question asked about other clinicians, "How are their skills?" Skills are important because they represent our ability to care for patients effectively and therefore are a reflection of the EMS profession.

As a student, you work hard developing your skills in order to be a trained EMS professional. Hours are spent developing accurate assessment skills. Spinal immobilization is rehearsed so many times that you could perform an immobilization in your sleep. IV skills are practiced over and over until "one stick" becomes your middle name. You become so rigid in perfecting splinting skills that your peers consider you as rigid as the splint itself.

The PCR is the permanent legal record that summarizes everything about the patient, care, and transport, from demographic information to assessment, interventions, and patient disposition. Whether accurate or inaccurate, complete or incomplete, the documentation preserves the care and transport only as you record it. It can be stated that documentation is the greatest skill because it permanently records all the other skills you perform and authenticates whether or not these other skills were performed correctly. What's in it for you? Your documentation is a reflection of how well you perform your skills as an EMS professional.

Protecting Your Career

For some of you, your career will rest upon your ability to produce quality documentation. Documentation is the skill that may be called upon to save your professional life. You may be involved in a civil court case that could negatively affect or even end your career, or you may be called to testify in a criminal case in which a lawyer will challenge your credibility by attacking your documentation. Your license may be held in the balance due to an alleged treatment error, and the quality and accuracy of your PCR documentation either will keep the card in your wallet or will put it in the scrapbook.

Through documentation, you hold the power to protect what you have worked so long and so hard for. Through documentation, you also have the power to erase the good that you have done for your patients. It is through documentation that you can damage your career simply by not making the commitment to protect yourself. Remember the following:

- If you didn't document a full assessment, you didn't complete a full assessment. Therefore, your assessment skills are poor.
- If you didn't document the IV flow rate, it could be assumed that you always infuse at "wide open." Therefore, your IV skills are poor.
- If you didn't document that you immobilized the patient, then you didn't perform the skill. Therefore, your immobilization skills are poor.
- If you didn't document that you secured the airway in the proper manner, then you didn't perform the skill. Therefore, your airway skills are poor.

Failing to attain proficiency in documentation is like providing spinal immobilization without a cervical collar or defibrillating without electricity. Mastering the skill of EMS documentation is an investment in your career. It's the one skill you perform for yourself.

Your Quality of Life

The quality of EMS documentation ultimately affects your paycheck. EMS systems are paid based upon the services provided. If EMS organizations fail to receive the reimbursement that is appropriately due, what are the chances that you'll see a raise or replacement of the outdated equipment you have all been complaining about? Documentation impacts not only your company's bottom line, but yours as well. If your EMS system consistently receives appropriate reimbursement for the services you and your peers are providing, your company will be healthier financially. When your company is financially healthy, you have a better chance of having better equipment as well as a better rate of pay. While documentation is certainly not the only determining factor affecting the financing of EMS, it is an important part of EMS reimbursement, and a tie exists between your documentation and your quality of life.

Summary: Return to Case Study

Let's return to the case study at the beginning of the chapter. There are two possible ways that this call can end.

1. In a hurry to return to the station you hastily complete your PCR in just a few minutes. You drop off a copy at the Emergency Department unit coordinator's desk on your way out the door. It's incomplete and leaves quite a bit to be desired in the legibility department, but your attitude is "I don't get paid the big bucks to do the billing department's job too, and the game is about to start."

As you step into the front of the ambulance, your partner comments: "Another PCR slam dunk. That one was a record. I like working with a guy who can get it done in less than five minutes. Sometimes I sit out here for almost a half hour waiting for my partner to finish a simple form." You reply: "Just enough to CYA, that's my rule. Step on it—we can be back to the station in time for kickoff."

2. While your partner is taking care of the truck, you find a seat in the report room and sit down to finish your PCR. You take out the recorder you used to dictate treatment times and start to complete the PCR. Beginning with the patient's demographic information, you meticulously document the Patient Care Report.

Paid or volunteer and at whatever level of training, attaining proficiency in documentation is one of the foremost investments you can make in your career in EMS. EMS documentation will play a leading role in defining and advancing the EMS profession.

CHAPTER REVIEW

Review Questions

Please refer to Answers to Chapter Review Questions at the back of this book.

1. List the three major challenges facing EMS today. Describe how the EMS professional can make an impact in meeting these challenges through the PCR.

2. Describe how PCR documentation reflects upon you as an EMS professional.

3. Why is documentation an EMS skill? Describe the importance of attaining proficiency in documentation skills to your EMS career.

4. Describe how developing proficiency in documentation protects the EMS professional's career.

5. Evaluate the PCR in Figure 1-2.

- What are your impressions of this documentation?

- How would you describe the patient care based upon what has been documented?

- How does this PCR represent EMS?

EMS Documentation PCR Example

Incident Location: *300 Main Street, Hasketa, OK*
Report Number: **2006000145**
Incident Date: *1/15/2007*
Receiving Hospital: *MRMC*

Last Name *Smith*	First Name *John*	Middle *B.*

Mailing/Home Address: *300 Main St.*

City: *Hasketa*	County *Carver*	State: *OK*	Zip: *74000*

SSN: *UTO*

Age: *51*	Date of Birth: *7/15/56*

Phone: *(918) 555-7777*

Chief Complaint: *Chest Pain*

Current Medications: *Nitro*

Allergies: *Unknown*

Patient Found: *Supine*

Stretcher Necessary? *Yes*   Reason: *Chest Pain*

TIME	PULSE	RESP	BP	LOC	SAT	EKG
	88	*20*	*110/70*	*X4*	*95%*	*NSR*
			/			
			/			

TREATMENT:	RESPONSE:
1. *NTG X3* 2. 3. 4.	

Narrative:
Pt. found seated in chair @ res c/o CP X 1hr. O2, monitor, IV started. NTG for CP. Transport to MRMC w/o incident.

Patient Signature: *UTO*

Crew 1: *255*	Crew 2: *321*

FIGURE 1-2
PCR Example

Critical Thinking

Please refer to Answers to Critical Thinking Discussion Exercises at the back of this book.

1. How does PCR documentation represent EMS as a profession?

2. The average paramedic spends 1,000 to 1,300 hours training. Of these 1,000 to 1,300 hours, it could be estimated that about 5 to 10 hours are spent in documentation skill training. Describe how this impacts documentation skill development.

Action Plan

Everything in EMS care begins with assessment.

1. What are your attitudes regarding documentation?

2. What would you consider to be your current level of documentation proficiency?

3. Regardless of "where you are responding from," make a commitment to becoming proficient in the skill of documentation.

Practice Exercises

Let's begin with discovering how you currently document key elements of an EMS call. This will serve as a baseline assessment of your documentation skill and will provide a means for you to evaluate your progress as we move through this discussion.

1. In the narrative space below, record the manner in which you typically document assessment of a patient with chest pain.

Narrative Snapshot

(Levels: Emergency Medical Responder, EMT, Advanced EMT, and Paramedic)

2. In the narrative space below, record the manner in which you typically document spinal immobilization.

Narrative Snapshot

(Levels: EMT, Advanced EMT, and Paramedic)

3. In the narrative space below, record the manner in which you typically document endotracheal intubation.

Narrative Snapshot

(Level: Paramedic)

CHAPTER 2

Documentation and Professionalism

Key Ideas

Upon completion of this chapter, you should know that:

- The definition and advancement of the EMS profession is the greatest challenge facing EMS today.

- Understanding how a profession is defined is essential to the advancement of EMS.

- Documentation plays a central role in defining the EMS profession.

- The history of EMS can be seen in the evolution of the Patient Care Report (PCR).

- Quality improvement and EMS research will advance the EMS profession, but they are dependent upon accurate documentation in the PCR.

- Attaining proficiency in documentation will advance EMS professionalism.

FIGURE 2-1
(Courtesy Acadian Ambulance Service, Lafayette, LA)

Return to the case study in Chapter 1. Unfortunately, you made the wrong choice. The Patient Care Report (PCR) was completed hastily in an effort to get back to the station. You didn't realize it at the time, but this decision will have lasting consequences.

While you are watching the game back at the station, the medical control physician at the Emergency Department (ED) is meticulously completing her documentation. Wanting to cover every base, the physician glances at the chest X-ray, and to her alarm she notices that the endotracheal tube is in the right main stem bronchus. The physician asks the charge nurse: "Do you have the EMS Patient Care Report on the trauma code?" The RN locates the PCR in the nurse's documentation room where the receiving RN has just finished the ED flow sheet. Making reference "to another masterpiece from the EMT named Bubba," the RN hands it to the charge nurse. The medical control physician reviews the PCR and immediately calls the on-duty EMS supervisor to report a patient care incident.

The next day, the physician makes a telephone call to the medical director of your EMS service. The Emergency Department physician explains there is a significant problem with "one of your paramedics." Finding your name and employee number at the bottom of the PCR, she relates receiving a patient from you that had been improperly intubated and your care might have contributed to the patient's poor outcome. Concerned, your medical director says: "The documentation will tell the story. Fax the PCR over to me." Within minutes your medical director is reviewing your PCR and immediately contacts the quality manager to initiate a formal investigation (see Figure 2-2). After the investigation you are placed on "performance review" for the next six months as the medical director has determined your care posed significant liability risk.

Questions

Evaluate the Patient Care Report in Figure 2-2. Please refer to Answers to Case Study Questions at the back of this book.

1. Based only on the documentation, how would you evaluate the patient care?

2. Describe the link between quality of care and quality of documentation.

3. How does the documentation in the PCR in Figure 2-2 reflect on the EMS provider?

4. The EMS provider in the case study is in trouble. Could it have been avoided?

EMS Documentation PCR Example	
Incident Location: *Hwy 9 and County 15* Report Number: *20060001736* Incident Date: *2/25/2007* Receiving Hospital: *County*	

Last Name *Doe*	First Name *John*	Middle

Mailing/Home Address: *unknown*			
City: *UTO*	County:	State:	Zip:

SSN: --------------

Age: approx *40s*	Date of Birth: *UTO*

Phone:

Billing Information:
Name/Company

Insurance Company Address: *See Hospital Records*
Group/ID Number:

Medicare #:   Medicaid #:
Self-Pay: Miles:

Chief Complaint: *trauma*

Current Medications: *unknown*

Allergies: *unknown*

Patient Found: *in vehicle*

Stretcher Necessary? *trauma code*   Reason:

TIME	PULSE	RESP	BP	LOC	SAT	EKG
scene	*80/40*	*rapid*	*120*	▼		*sinus*
en route	*80/40*	*rapid*	*120*			*sinus*

TREATMENT:	RESPONSE:
1. *spinal immobilization* 2. *O2/IV/Monitor* 3. *ET Tube* 4.	*Emergency Medical Responders*

Narrative:
Trauma. Followed Trauma/ACLS protocols. Care turned over to hospital staff.

Patient Signature: *unable*

Crew 1: *P. Crew*	Crew 2: *T. Smythe*

FIGURE 2-2
PCR Example

Introduction

What Is EMS?

Is EMS health care, public safety, or public service? Perhaps you've wondered how you "fit in" to the overall health care system as an EMS professional. Does the term *EMS professional* seem inconsistent with your daily work experience? Are you treated as a professional by other

Table 2-1	EMS Associations
National Registry of EMTs	American Ambulance Association
National Association of EMTs	International Association of Flight Paramedics
National Association of EMS Educators	International Association of Firefighters
National Association of State EMS Directors	International Association of Fire Chiefs
National Association of Female Paramedics	National Council of State EMS Training Coordinators
National EMS Management Association	National EMS Pilots Association
National Flight Paramedics Association	National Fire Protection Association
National Emergency Management Association	National Emergency Medicine Association
National Association of EMS Physicians	International Association of Emergency Managers
National Association for Search and Rescue	National Association of Fleet Managers

members of the health care team at the hospital, clinic, nursing home, or urgent care center? Do you sense that physicians and nurses understand your capabilities? Have you ever felt that the respect you have on the street erodes by the time you are in the hospital corridors?

Who Speaks for the EMS Profession?

Many professional associations are involved in EMS today. Table 2-1 is a partial listing of these organizations. Each makes a contribution to the EMS industry. However, with many voices speaking from various interests within the industry, no single unifying voice speaks for the EMS profession. The American Nurses Association "speaks" on behalf of the registered nurse, the American Medical Association "speaks" on behalf of the physician, but who is speaking for the EMS professional?

Who Regulates the EMS Profession?

Although many organizations are involved in regulating EMS, it has yet to be determined which government agency is ultimately responsible. Is it the U.S. Department of Transportation (DOT) or the National Highway Traffic Safety Administration (NHTSA) or the Health Resources and Services Administration (HRSA)? How do state EMS regulations and medical direction fit in? Perhaps in moving from state to state you have experienced frustration because of lack of licensure reciprocity and the differences in continuing education and refresher requirements. With currently 44 different levels of EMS practitioners across the country, inconsistencies between agencies can cause confusion in licensure, recertification, and continuing education requirements.

PCR documentation is a reflection of the confusion existing in EMS today:

- With no universal documentation format, PCRs can vary from service to service, county to county, or state to state.
- Documentation content requirements of individual EMS providers often vary greatly.
- Documentation formats are as varied as EMS uniforms.

 Professionalism in EMS is the display of the defining characteristics of the EMS profession.

Professionalism and documentation are inseparable. Any profession is defined, established, recognized, and advanced through the recording of its unique professional activities. Transcriptions of procedure and examination notes reflect the medical profession. Legal briefs, depositions, and court decisions reflect the legal profession. Care plans and the nursing diagnosis reflect the nursing profession. The Patient Care Report reflects the EMS profession. Professionalism in EMS is the display of the defining characteristics of the EMS profession.

KEY TERMS

Note: Page numbers indicate where the following key terms and definitions first appear.

autonomy (p. 17)   **dataset (p. 24)**

Defining the EMS Profession

A profession can be defined as follows: "a calling requiring specialized knowledge and other long and intensive academic preparation."[1]

Defining the EMS profession provides a template, or a pattern, by which EMS professionals serve the public. Just as when the public uses the services of a physician or attorney, there is an expectation of what will be received based upon established professional standards; when an EMS provider responds to an emergency, the public should receive the same excellence in care, no matter who shows up in the uniform. Figure 2-3 lists attributes of the emerging EMS profession.

autonomy:
The professional ability to work independently from others.

National identity: A profession has an established national identity with a unified national association that leads and speaks for members of the profession.

Defined leadership: A profession has well-defined leadership from within the profession. For the EMS profession to advance, leadership must come from a single professional association that leads and speaks for members of the profession.

Specialized EMS knowledge and education: A profession has specialized knowledge and education. EMS knowledge and education is unique to health care and therefore provides members of the EMS profession a platform to display the uniqueness of the profession through daily EMS practice.

Autonomy: A profession has **autonomy**. A profession is autonomous when its members provide direction for its behaviors and practice. One important aspect of autonomy is financial awareness. EMS professionals, down to the street level, must have awareness of the business of EMS and the financial challenges associated with the profession.

High ethical standards: A profession demands high ethical standards of its membership. As EMS emerges as a profession, members will have ownership for establishing and enforcing high ethical standards.

FIGURE 2-3
Attributes of the Emerging EMS Profession

[1]*Merriam-Webster's Collegiate Dictionary,* 11th ed. (Springfield, MA: Merriam-Webster, 2003), p. 991.

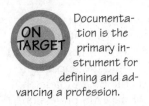

ON TARGET

Documentation is the primary instrument for defining and advancing a profession.

The EMS professional advances the EMS profession one patient encounter at a time, as these attributes of the EMS profession are displayed to the public. The Patient Care Report, as the representative document of the profession, records our professional activities. Because PCR documentation is the only representation of the EMS profession to those outside EMS, no professional activity represents EMS more than PCR documentation.

The History of the EMS Profession and Documentation

Having an historical perspective of the evolution of EMS is vital to the advancement of professionalism. In order to understand the importance of documentation in the advancement of the EMS profession, the history of EMS must be understood. The evolution of EMS can be seen in the Patient Care Report. Imagine a time when there were no organized EMS systems or no 911 service. In order to summon an ambulance one had to find the phone number for the local funeral home or volunteer rescue squad (see Figure 2-4).

Today placing a call for EMS typically activates a well-choreographed emergency response system. From the call to the 911 communications center, where pre-arrival instructions are given and multi-agency resources are coordinated, to the arrival of the patient at the ED, all that goes into an EMS incident is the result of the evolution of EMS from the local funeral home to a multifaceted emergency medical services system.

Emergency Medical Services in the United States can be traced back to the first volunteer rescue squad, started by Julian Stanley Wises in Roanoke, Virginia, in the 1920s. Volunteer rescue squads and funeral homes provided "ambulance service" throughout the country during this period of time. Modern EMS became visible in the 1950s as a result of the development of CPR, which later led to the development of ACLS and the Chain of Survival. The advance of CPR provided the first formal EMS training for many ambulance rescue squads.

Ambulance services were generally a side business of the local funeral home and consisted of an ambulance and one or two attendants. Ambulances were most often Cadillacs, familiar equipment to this industry bearing resemblance to the primary vehicle in its fleet, the hearse (see Figure 2-5).

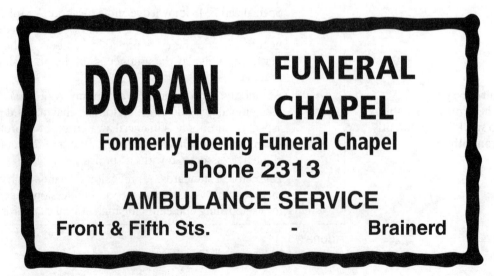

DORAN FUNERAL CHAPEL
Formerly Hoenig Funeral Chapel
Phone 2313
AMBULANCE SERVICE
Front & Fifth Sts. - **Brainerd**

FIGURE 2-4
Funeral Home Advertisement

FIGURE 2-5
Cadillac Ambulance (Courtesy Tony Karsnia)

Because these old ambulances served only a transportation function, there wasn't a need for any room to work in the back, just a place for the two-person cot. The "attendants" usually had no formal training and both attendants often rode in the front seat, leaving the patient to wait for care at the hospital. During this period, at least 50 percent of our nation's ambulance services were provided by 12,000 of the nation's morticians.[2]

The Early Years—EMS in the 1960s and 1970s

A number of individuals could be considered founding fathers of modern EMS. One of the most important would be President Lyndon Baines Johnson. His vision of the "Great Society," in which government would provide solutions to the societal issues of poverty and health care, would ultimately result in the establishment of the first EMS systems. President Johnson signed the Highway Safety Act of 1966, the first legislation relating to Emergency Medical Services (see Table 2-2).

Table 2-2 Highway Safety Act of 1966	
Highway Safety Act of 1966	**EMS Implications**
Established Department of Transportation	Developed standards for training
Mandated improvement of EMS systems	Provided $142 million for EMS system development via DOT
Required states to develop state EMS systems	Resulted in the first national EMT-A curriculum
Promoted state EMS legislation	Encouraged the formation of state EMS offices

[2]Digby Diehl, "The Emergency Medical Services Program," Robert Wood Johnson Foundation (2000). Retrieved March 19, 2007, from www.rwjf.org/files/publication/books/2000/chapter-10.html.

Table 2-3 Recommendations of the "White Paper"—*Accidental Death and Disability: The Neglected Disease of Modern Society*

Expansion of Basic and Advanced First Aid Training to the General Public	Preparation of Formal Training Courses with Textbooks and Training Equipment
Establishment of standards for ambulance design	Establishment of standards for ambulance equipment and supplies
Establishment of standard qualifications for ambulance personnel	Establishment of state EMS regulations for ambulance services
Establishment of methods for providing appropriate levels of ambulance service and integrating these services into health and public safety	Establishment of radio frequencies dedicated to ambulance services in order to provide radio contact between hospitals and the public service sectors
Promotion of the idea of a single nationwide telephone number for emergency services	Pilot programs to explore the feasibility of helicopter ambulance service

The Highway Safety Act of 1966 was the legislative response to the "White Paper"—*Accidental Death and Disability: The Neglected Disease of Modern Society*. Published by the National Academy of Sciences, the White Paper identified traumatic injury as "the neglected disease of modern society" and revealed the absence of medical care outside the walls of the nation's hospitals. The White Paper served as a wake-up call leading politicians, physicians, and policy makers to address the nation's inability to care for the sick and injured outside hospitals (see Table 2-3).

Because EMS in the 1960s served the function of emergency transportation, its documentation reflected only the transportation aspects of each ambulance call. Prior to Patient Care Reports or "trip sheets," many ambulance services only recorded their calls on a log kept by the dispatcher, who was often the same person that responded. Documentation centered on the basic elements of transportation: name, address, pickup location, hospital, mileage, problem, and charge amount. The simple form was primarily a means to generate a bill for the transportation provided to the patient. Because transportation was the primary function, the ambulance was the star, and the attendants were the supporting cast.

By the 1970s, EMS was experiencing significant advancement as a result of the White Paper and the Highway Safety Act of 1966. While politicians were responding to the White Paper, the public remained in the dark until the television show *Emergency* debuted in 1971. Each week Johnny Gage and Roy Desoto responded from Station 51 to homes across America. *Emergency* not only provided the nation its first educational exposure to Emergency Medical Services but also had a profound effect on the development of EMS.

The Emergency Medical Services Systems Act of 1973 defined EMS systems and provided federal funding for the development of EMS systems. Championed by Congressman Gerald R. Ford of Michigan, this legislation established the Division of Emergency Medical Services at the Department of Health, Education and Welfare (now known as the Department of Health and Human Services) and provided $300 million for EMS research, operations, and expansion. In 1974, President Gerald Ford signed the first EMS Week proclamation.

During the 1970s, transportation met medicine, and the evolution into emergency care in the field began with the creation of the Emergency Medical Technician—Ambulance (EMT-A) program. The 1970s were EMS's golden hour era of "load and go" and "scoop and run" as the first certified EMTs hit the streets. Once patient care began in the field, medical documentation was required. The ambulance log was no longer sufficient as EMTs provided basic patient care along with transportation, and the individual trip sheet, one for each patient encounter, was required. Although these early trip sheets varied greatly among ambulance providers, they contained the same essential components. Dispatch logs began recording key dispatch information and the trip sheet recorded the condition and care of the patient. Vital signs, oxygen, CPR, splinting, and spinal immobilization were recorded as well as a probable diagnosis. See Figure 2-6 for an example of a 1970s trip sheet. EMS documentation in the 1970s reflected EMS at that time—the emerging technician.

Acadian Ambulance Service, Inc.

P. O. BOX 52888, O.C.S. · LAFAYETTE, LOUISIANA 70505
PHONE (318) 232-1511 OR DIAL TOLL FREE 800-252-5522

Date ________ Time ________ A.M / P.M Type Call ________ SCR# ________

Name ______________________________ ________ Age ☐☐☐

________ Medicare   # ________ Medicare   # ________ EMS

From ______________________________

To ______________________________

Illness or Injury ________________ Doctor ________________

☐ Member #: ________________ ☐ Non Member

________ Local Emergency   ________ Stand-by Services

________ Local Transfer   ________ Oxygen

________ Night Charge   ________ EKG Monitor

________ Offshore Emergency   ________ IV Set Up

________ Extended Transfer   ________ Additional IV Fluids

________ Emer. Ext. Transfer   ________ Other

Total Charge $ ________ ☐ WC   Amount Paid $ ________

Charge To ______________________________

Address ______________________________

City & State ________________________ Zip ________

Phone# ________________ Relation ________________

Employed By ________________ Phone# ________________

Business Address ______________________________

Alt. Contact ________________ Relation ________________

Address ________________ Phone ________________

City & State ________________________ Zip ________

Driver ________ Technician(s) ________ Unit ________

I agree to pay the amount within 30 days from date. I further agree that in the event this amount is not paid within 30 days, that I shall be liable for 18% annual interest on the balance until paid. Furthermore, In the event it becomes necessary to place this account in the hands of a Credit Bureau and/or a Collection Agency, I agree to pay any collection fees on the balance due and owing.

Signed X ______________________________

Special Authorization ________________ ________ Parish   24883

FIGURE 2-6
Example of 1970s Run Report (Courtesy Acadian Ambulance Service, Lafayette, LA)

FIGURE 2-7
Example of 1980s Patient Care Report (Courtesy Acadian Ambulance Service, Lafayette, LA)

EMERGENCY CHARGE TICKET
Acadian Ambulance Service, Inc.
P. O. BOX 92970 • LAFAYETTE, LOUISIANA 70509-2970
BUSINESS (318) 261-1522 • OR 1-800-523-2823
PRESS HARD

90152

PERISH				
DATE		ORIG. SCR		
UNIT #	SIG	SCR		

PATIENT'S NAME — Last — First — Middle — AGE — RACE — SEX — DATE OF BIRTH
ADDRESS — CITY — STATE — ZIP
FROM — TO — PATIENT'S PHONE ( )
GUARANTOR'S NAME: Last — First — Middle — RELATIONSHIP — PHONE ( )
ADDRESS — CITY — STATE — ZIP
ALTERNATE CONTACT — RELATIONSHIP — PHONE ( )
ADDRESS — CITY — STATE — ZIP
EMPLOYER (If retired - from what co.)
ADDRESS — CITY — STATE — ZIP — PHONE ( )

SERVICES PROVIDED
☐ BLS Emergency ☐ Return Trip ☐ Oxygen Set Up ☐ Suction ☐ Monitor ☐ Additional Fluids
☐ ALS Emergency ☐ Night ☐ C-Collar ☐ OB Kit ☐ Monitor w/Defib. ☐ Visidex II
☐ Transfer ☐ Mileage ☐ Mast ☐ Poison Kit ☐ E.O.A. ☐ Drugs
☐ Physician Fee ☐ Sterile H₂O ☐ Burn Sheet ☐ Intubation ☐
TOTAL CHG. — AMT. PAID — BAL. DUE — REC'D BY ☐ Cardbd Splint ☐ I.V. Insertion
☐ Disaster Bag ☐ I.V. Set Up

☐ MEM
☐ NONMEM. ☐ W.C. ☐ Other
☐ DR. LIC. _____ STATE: _____
☐ S.S.
☐ MEDICARE
☐ MEDICAID
MTC Left: ☐ No ☐ Yes, Where:

PAYMENT AUTHORIZATION & INFORMATION RELEASE
I authorize payment directly to Acadian Ambulance Service, Inc. of the ambulance benefits otherwise payable to me. I authorize any holder of medical or other information about me to release to my ambulance benefits provider, including the Social Security Admin., or its intermediates or carriers if I have Medicare coverage, any information needed for this or related claims. I permit a copy of this authorization to be used in place of the original. I understand I am financially responsible to AASI for charges not covered by this Authorization. I am liable for a finance charge of 1% per month on the unpaid balance (12% annual rate) until paid in full. If collection of my past due account is pursued by suit or otherwise I agree to pay all collection costs, including 25% attorney fees.

SIGNED: _____
DATE: _____ GURANTOR: _____
☐ The patient was transported to this facility at the request of
_____ (Relationship to Patient)
Signed _____

MEDICAL REASON FOR TRANSPORT:
(Impression, Why Stretcher Needed, Why Hospital To Hospital)

FULL NAME OF RECEIVING/TREATING PHYSICIAN
☐SPECIAL AUTHORIZATION
Insurance Company: Group ☐ Individual ☐
Address:
Policy #: _____ Group #: _____
Name Of Insured: _____ S.S.#

| SUSPECTED ILLNESS/INJURY, ADMIT/DISCHARGE DIAGNOSIS | INJURY PROBABLE CAUSE |

MD ORDERING

DESTINATION	HOSP. TREATMENT	TRAMA/APGAR	PERSONS RIDING
☐ HOSP.	☐ IV FLUID	TIME/SCORE	☐ MD
☐ CLOSEST HOSP.	☐ MEDS	/	☐ NURSE
	☐ NG TUBE	/	☐ FAMILY
	☐ ET TUBE	/	
☐	☐		☐

C/O — TIMES — VITALS | BP — P | R
PMH
ALLERGIES
MEDS
HPI
PE _____ KG'S (#/2.2)

DISPATCH TIMES
	EN ROUTE		AR. SCENE
HOSP.	LV. SCENE		AR. HOSP.
TIME	PERSONS RIDING	TIMES	BLS/ALS TX

MIN. DELAY FOR EXTRICATION		ASSISTANCE P.T.A.		
MEDICS	1	2	3	4	5
REG.#	#	#	#	#	#

ADDITIONAL COMMENTS ON BACK
AASI (4-87) (301)

The Transition Years—EMS in the 1980s and 1990s

EMS in the 1980s was characterized by the established technical role of the EMS provider. In the 1980s EMS emerged as a health care business as federal funding eroded. Ambulances, equipment, and supplies had to be purchased and payrolls for EMS staff had to be met. Patient Care Reports in the 1980s, comprised of open fields and check boxes, reflected the expanding clinical capabilities with dedicated space to document advanced airway procedures, medication administration, and cardiac arrest management. EMS professionals' signatures and license numbers (or National Registry numbers) became a requirement as the Patient Care Report was recognized as a legal document. EMS documentation in the 1980s reflected EMS at that time—the established technician. See Figure 2-7 for an example of a 1980s Patient Care Report.

EMS in the 1990s was characterized by the challenges associated with an evolving profession. State EMS agencies became well established and more involved in the patient care activities of EMS providers in their states. Until the 1990s, the Patient Care Report had been the unique creation of individual providers to suit their own business needs. In the 1990s EMS leaders slowly began to realize the importance of research to the advancement of EMS. Many states began using universal state EMS forms. In order for valid research to take place, a uniform documentation tool was required. PCRs in the 1990s typically used a system of check boxes or circles to facilitate the extraction and collection of data while retaining open fields for demographic information, assessment findings, and treatment. The narrative section remained for the EMS encounter to be summarized.

The financial demands on EMS systems increased in the 1990s as DRGs became universal across health care. EMS providers were required to establish the medical necessity of EMS services. In response, progressive EMS providers began requiring medical diagnoses, past medical history, and bed confinement/ambulation status in order to document medical necessity for transport. EMS documentation in the 1990s reflected EMS at that time—the established clinician.

EMS Today

The last 40 years have brought EMS from serving only as transportation to providing transportation with medical care, to now being an integral part of community health care (see Figure 2-8). EMS today continues in a state of transition as it seeks to establish a professional identity distinct from public safety, medicine, and nursing.

EMS documentation, through the Patient Care Report, is one of the primary means by which the EMS profession communicates many of its attributes: specialized knowledge,

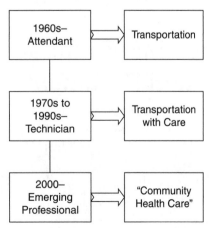

FIGURE 2-8
The Evolution of the EMS Professional

high ethical standards, and autonomy. A profession's means of documentation represents its autonomy over its professional activities. For example, nursing education emphasizes the nursing diagnosis and care plan. Although nursing students may question their place in the reality of practice, the nursing diagnosis and care plan have become symbols of the nursing profession and represent the specialized knowledge and autonomy of the registered nurse as expressed in documentation. In doing so, professional nursing has moved away from physician-centered (physician's orders) documentation to documentation centered on the profession of nursing. Similarly, the EMS Patient Care Report can reflect the unique activities of the EMS practice and can play an important role in advancing the EMS profession. Tremendous demands are now placed upon EMS documentation: today's Patient Care Report is a clinical, legal, financial, and research document, and proficiency in documentation is essential to meeting these demands and advancing the profession.

Advancing the EMS Profession—Research and Documentation

EMS research began with the passage of the EMS Systems Act of 1973, which allocated funding not only for development of EMS systems but also for research in emergency techniques, methods, devices, and delivery. Because early EMS research was lacking in science, methodology, and funding, it had limited value. EMS research is no longer an option, and it serves two valuable functions. First, research provides the justification for EMS by questioning and evaluating current practices. Second, it advances the EMS profession. Back in the 1970s and early 1980s when EMS personnel were the "eyes and ears of the physician," it may have sounded great, but was it really? If EMS providers were the eyes and ears, who was the brain? At that time the physician in the ED was the brain and the EMS providers were only technicians, unable to think critically and arrive at clinical decisions on their own. Therefore, a physician was needed to make the decision to transport a patient, start an IV of D5W, or diagnose a third-degree heart block via telemetry. It wasn't until research proved that EMS providers could assess, make a preliminary diagnosis, and render appropriate treatment that EMS roles began to change (see Figure 2-9).

dataset:
Groups of data that describe every aspect of an EMS event.

Data collection has become the primary focus of research, prompted by the realization that research requires data. To this end, the National Highway Traffic Safety Administration (NHTSA) has established a system for EMS data collection. In 1994, NHTSA released the first version of the Uniform PreHospital EMS Datasets. **Datasets** are subsets of information obtained for every patient encounter. Datasets will be discussed in Chapter 6. The National Emergency Medical Services Information System (NEMSIS) project was initiated to serve as the collection center of standardized EMS data. The focus on data collection for EMS research places a tremendous demand on EMS professionals for accurate documentation.

Advancing the EMS Profession—Quality Management and Documentation

Quality improvement began in the 1980s as quality assurance (QA). As EMS systems developed, medical directors and administrators realized the importance of assuring quality of care. Initial QA programs usually involved utilizing supervisors to review Patient Care

FIGURE 2-9
The Continuum of Change—EMS Research

Reports to verify that care was in accordance with protocols and that documentation was complete.

In the 1990s, quality assurance became quality improvement (QI). It was deemed insufficient simply to assure quality; quality must be improved. Quality improvement became more sophisticated as organizations developed formal quality programs. A typical quality program at this time not only focused on completeness of documentation but also measured documented care against standards of care. Many organizations began programs that established measurable standards of quality and performance benchmarks. Today, continuous quality improvement (CQI) is an integral part of most EMS systems. Unlike the retrospective approaches of quality assurance and quality improvement, which centered on deficiencies, quality management takes a prospective approach to quality and identifies best practices for the entire EMS system.

In our case study at the beginning of the chapter, was the problem with the patient care or with the documentation of the patient care? The perceived quality of patient care is closely linked to the quality of documentation. Once the call has been completed, the PCR documentation permanently represents the care given. It is the expectation of everyone, from your employer to the lawyer who may be representing one of your former patients, that documentation equals patient care. Therefore, quality management and documentation will be forever linked. The Patient Care Report is the most valuable tool in evaluating and improving the quality of care. Because documentation quality and the quality of patient care are inseparable, PCR documentation will always be central to the quality management of patient care activities. Quality management is a great asset to EMS in attaining documentation proficiency. It is far better for deficiencies, revealed in documentation, to be brought to light by "friends" within the EMS organization than by "foes" outside the organization. Quality management advances the EMS profession by assuring that the highest level of patient care is delivered to all patients by evaluating patient care against measurable standards.

Summary: Return to Case Study

Once again, let's return to the case study. In review, you chose to document patient care that was not reflective of the care that you actually gave. You were lazy, putting recreation ahead of your responsibilities to your profession and your career. As a result of the choice that you made regarding documentation:

- Judgments have been made about you as a professional and, as a result, stereotypes have been reinforced.
- Your quality improvement manager has put you on a six-month review. The index of suspicion is now upon you. Every clinical move you make will be closely scrutinized.
- Unfortunately, in EMS it takes only a second to make an error in judgment. In the minds of the medical control physician, medical director, and supervisors, your patient care abilities are now in question. Once a question mark is placed over professional abilities, erasing it will be difficult.

The case study demonstrates the decision that is before every EMS provider each time a Patient Care Report is completed, and this decision can have dramatic effects on how you are viewed as a professional. Your Patient Care Report hasn't been reviewed by everyone, yet. It's now on the way to the billing department. To be continued.

History has shown the progression of EMS provider from attendant to technician to emerging EMS professional. Patient Care Report documentation represents many of the essential attributes of the EMS profession. Attaining proficiency in documentation is a great investment in the advancement of the EMS profession.

Review Questions

Please refer to Answers to Chapter Review Questions at the back of this book.

1. Why is it important for the EMS profession to be clearly defined?

2. List attributes of the emerging EMS profession.

3. What was the focus of documentation in the 1960s and 1970s?

4. Describe how PCR documentation changed in the 1980s and 1990s.

5. Describe the impact of the White Paper on the development of EMS.

6. Describe the impact of the Emergency Medical Services Systems Act of 1973 on the development of EMS.

7. Describe the relationship between PCR documentation and the advancement of the EMS profession through research.

8. Describe the relationship between PCR documentation and the advancement of the EMS profession through quality management.

Critical Thinking

Please refer to Answers to Critical Thinking Discussion Exercises at the end of this book.

1. Describe your current attitude toward quality management. How do you generally respond to negative feedback on your patient care, and how does this relate to proficiency in documentation?

2. What is the difference between a profession and a vocation? Does EMS qualify as a profession? Are EMS skills professional skills or technical ("trade") skills?

Action Plan

1. Review each PCR you complete for accuracy and completeness. Understand the ramifications of incomplete and inaccurate documentation.

2. As you complete your PCRs, ask yourself:
   - Does this documentation represent my best effort?

   - What does this documentation say about me as a professional?

   - What does this documentation say about the EMS profession?

3. Study the history of EMS.

4. Review some of your PCRs from last year. Look for problem areas. Are you getting sloppy? Are bad habits on the horizon? Are you improving in documentation? Make it a practice to assess your own documentation skill development periodically.

5. View episodes of the television program *Emergency*.

6. Read the *EMS Agenda for the Future*.[3] Did the *Agenda* define EMS as a system or as a profession?

Practice Exercises

1. Rewrite the Patient Care Report from the case study to your level of licensure. Use **Figure 2-10 for the exercise.**

[3]The *EMS Agenda for the Future* established a vision for the future of EMS as an integral component of the nation's health care system. The *Agenda* proposed and defined 14 attributes of EMS focusing only on the development of EMS as an agency (or system) and not as a health care profession. http://www.nhtsa.dot.gov/people/injury/ems/agenda/emsman.html.

EMS Documentation PCR Example						
Incident Location: Report Number: Incident Date: Receiving Hospital:						
Last Name		First Name			Middle	
Mailing/Home Address:						
City:		County:	State:		Zip:	
SSN:						
Age:			Date of Birth:			
Phone:						
Billing Information: Name/Company: Insurance Company Address: Group/ID Number: Medicare #: Medicaid #: Self-Pay: Miles:						
Chief Complaint:						
Current Medications:						
Allergies:						
Patient Found:						
Stretcher Necessary?				Reason:		

TIME	PULSE	RESP	BP	LOC	SAT	EKG
				▼		

TREATMENT:	RESPONSE:
1.	
2.	
3.	
4.	

Narrative:

Patient Signature:	
Crew 1:	Crew 2:

FIGURE 2-10
PCR Exercise

2. Evaluate the following documentation example (Figure 2-11a). Does this documentation example represent that professional patient care was delivered to the patient? Rewrite this example in order to better reflect the EMS profession (Figure 2-11b).

TIME	PULSE	RESP	BP	LOC	SAT	EKG
	140	16	90 /80	▼	90	Sinus
			/			

TREATMENT:				RESPONSE:		
1. O2						
2. IV						
3. Monitor						
4.						

FIGURE 2-11a
Documentation Example

TIME	PULSE	RESP	BP	LOC	SAT	EKG
			/			
			/			

TREATMENT:				RESPONSE:		
1.						
2.						
3.						
4.						

FIGURE 2-11b
Documentation Example

3. Evaluate the following narrative example (Figure 2-12a). Does this narrative example represent that professional patient care was delivered to the patient? Rewrite this example in order to better reflect the EMS profession (Figure 2-12b).

Narrative Snapshot
Patient spinally immobilized. Fall due to intoxication. Response delay as dispach sent wrong unit. Walked patient out of bar. Combative. This BLS call done by ALS crew.

FIGURE 2-12a
Narrative Example

Narrative Snapshot

FIGURE 2-12b
Narrative Example

4. Evaluate the following narrative example (Figure 2-13a). Does this narrative example represent that professional patient care was delivered to the patient? Rewrite this example in order to better reflect the EMS profession (Figure 2-13b).

Narrative Snapshot
Patient transported from hospital to nursing home.

FIGURE 2-13a
Narrative Example

Narrative Snapshot

FIGURE 2-13b
Narrative Example

CHAPTER 3

Documentation and the Financing of EMS

Key Ideas

Upon completion of this chapter, you should know that:

* The financing of EMS is one of the profession's foremost challenges.

* The patient care and financial aspects of documentation are equally important.

* The EMS professional must have an awareness regarding how EMS is financed.

* The Patient Care Report is the primary financial tool of EMS.

* Documentation and reimbursement are inseparable.

* Federal Health Care Program reimbursement is vitally important to EMS, and the EMS professional must have a basic understanding of how reimbursement relates to PCR documentation.

* Attaining proficiency in documentation will have a positive impact on the financing of EMS.

FIGURE 3-1
The Signing of the Medicare Bill (Courtesy of the Lyndon Baines Johnson Library)

Several mornings later your PCR arrives in the finance department. Even though finance is located at the other end of your station, it is a world away from daily EMS operations. Finance is composed of accounts receivable, payroll, collections, and the billing department. The billing department processes approximately 30,000 ambulance transports each year, turning transports into "claims" for payment. Your PCR is on the top of the billing specialist's stack of 200 reports this morning. Beginning the review, she sighs: "What a way to start a Monday. Sometimes I feel like a detective having to figure out what some of these medics actually did on these calls, and this guy is on the top of the list of the challenged." Despite your lack of documentation and the challenge of deciphering your handwriting, a level of service and condition code are assigned, and your call is electronically billed to Medicare.

At Medicare, the claim is processed, and based upon the level of service that your company billed, a "prepay medical review" is authorized to establish whether the ambulance transport was medically necessary. Therefore, Medicare sends a letter to your billing department requesting a copy of the PCR. Once your PCR arrives back at the Medicare carrier, a medical review nurse meticulously reviews it and comments to the analyst in the next cubicle: "This ambulance documentation is very poor. I want to pay this claim but I can't tell what they did." Unable to spend any additional time reading your PCR, the nurse analyst denies the claim, issuing the denial "documentation does not support medical necessity." Within a week the finance department is notified that the claim has been denied.

Questions

Please refer to Answers to Case Study Questions at the back of this Exercises book.
1. What made the EMS services in the case study nonpayable?

2. What role does PCR documentation have in the financing of EMS?

Introduction

We often think of health care only in terms of patient care. The clinical world of ambulances, helicopters, spinal immobilization equipment, cardiac monitors, and IV equipment often seems removed from the financial, or business, side of EMS. However, the passion we have for everything in EMS is totally dependent upon the business of health care. It take tremendous amounts of money to purchase and stock ambulances, keep the fuel tanks full, pay the insurance, and keep paychecks coming on time.

The financing of Emergency Medical Services is one of the foremost challenges facing the EMS profession today. As discussed in the previous chapter, an important attribute of professionalism is financial awareness. Today's EMS professional has the responsibility to be financially aware of how EMS is funded; in other words, of where the money comes from and how the bills are paid.

The financial aspects of documentation are almost always the most controversial. One reason for this is that EMS professionals have historically lacked understanding of how EMS services are reimbursed and how EMS pays its bills. Our counterparts in medicine would consider it unthinkable that documentation served only a clinical function. Physicians have

Table 3-1	EMS Financial Fallacies

- The Patient Care Report is only a clinical document.
- "Billing" and EMS operations must be separate.
- The financial aspects of documentation should be left to the billing department.
- All problems relating to EMS finances can be traced back to Congress, Medicare, and the Balanced Budget Act.
- If you teach EMS professionals about EMS finances and documentation, you are committing Medicare fraud.
- EMS finances and documentation are tricky and too complex for the average EMS professional to understand; therefore, these subjects should be avoided.
- Medicare denies ambulance claims at random and for no good reason.

a high degree of financial awareness and understand the importance of complete and accurate documentation so they receive appropriate payment for their services. If we are to attain the stature and benefits of professionalism, we must gain an accurate understanding of EMS finances, reimbursement, and the critical role documentation has in reimbursement. EMS professionals must first recognize possible errors regarding reimbursement and documentation. Table 3-1 lists a few of the more common financial fallacies that must be overcome.

In order to meet the financial challenges facing EMS effectively, we first must understand what the challenges are and how they have developed. This chapter presents an overview of the financial challenges facing EMS, how EMS is funded, how reimbursement is received, and the critical role of documentation in EMS financial processes.

KEY TERMS

Note: Page numbers indicate where the following key terms and definitions first appear.

Medicare fee schedule (p. 34)

Prospective Payment System (PPS) (p. 34)

fee for service reimbursement (p. 34)

Federal Health Care Program (p. 34)

subsidies (p. 34)

ICD9 (p. 35)

condition codes (p. 35)

procedure code (CPT/HCPCS Code) (p. 35)

The Financial Challenges of EMS

Most of us have a pretty good idea as to the condition of our personal finances. Anything less is irresponsible because we have financial obligations to meet. Overtime shifts may even be worked to meet these obligations. Yet most EMS professionals have little understanding of where the EMS profession is financially. It is crucial that the EMS professional have a basic understanding of EMS finances. Consider the following financial challenges facing EMS:

1. **The Cost of EMS:** Providing the nation with EMS is a multibillion-dollar effort each year. EMS is a costly enterprise. In 2002, Medicare paid nearly $3 billion for ambulance transports alone.[1]

[1]Department of Health and Human Services, Office of Inspector General (2006). *Medicare Payments for Ambulance Transports*. Retrieved August 7, 2007, from http://oig.hhs.gov/oei/reports/oei-05-02-00590

2. **Federal Health Care Program Payments:** Through the Balanced Budget Act of 1997, Congress mandated payment of EMS services according to a **Medicare fee schedule** based upon seven levels of service. The fee schedule predetermines payment, before 911 is even dialed, for each level of service.

3. **Medicare Prepay Medical Review:** Many Medicare carriers and insurance companies evaluate PCR documentation to determine wheather services meet medical necessity criteria prior to making payment.

4. **Diminishing Subsidies:** Governmental subsidies for EMS have declined as municipal budgets have tightened. Therefore, EMS systems that were once able to offset costs with subsidies are now finding themselves more dependent upon Medicare, Medicaid, and private insurance reimbursement.

5. **High Fixed Operational Costs:** Operating costs for Emergency Medical Services relate to readiness. EMS resources must be available to meet the demands of emergency and nonemergency transports. The total cost per year for operating the nation's public and private EMS systems is increasing as vehicle, equipment, labor, and administrative costs have increased while reimbursement from Medicare and Medicaid often falls below the operating costs.

Meeting these complex challenges will take the combined efforts of EMS leadership at the national, state, and local levels. However, every EMS professional can make a positive impact on the financial challenges facing EMS, one EMS incident at a time, through accurate PCR documentation.

EMS Funding—Past and Present

As EMS has evolved, its funding has changed. In the past, EMS systems relied upon funding from the federal government and reimbursement by Medicare, Medicaid, and private insurance companies. Prior to the implementation of the Diagnosis Related Groups (DRGs) in the 1980s, if an ambulance provider billed $500 for a transport, the provider would be paid $500 by Medicare and the private insurance companies. DRGs significantly changed the financing of EMS by ushering in what is known as the **Prospective Payment System (PPS).** This system predetermines what will be paid for a particular health care service. Compounding the problem, the Consolidated Omnibus Reconciliation Act, or COBRA, of 1985 ended federal financing of EMS.

Today, most EMS systems are financed in the same manner as other health care providers are funded through **fee for service reimbursement** from **Federal Health Care Program** reimbursement (Medicare, Medicaid) and private insurance companies. Some EMS systems are also supported by tax **subsidies** from county or municipal governments. With the majority of EMS financing directly associated to the actual EMS services that were provided, the Patient Care Report takes center stage. Therefore, the EMS professional must understand the relationship between PCR documentation and reimbursement.

Documentation and Reimbursement

Documentation and reimbursement are inseparable. This is true for the physician, physical therapist, psychiatrist, and dentist, and it is true for EMS. After the care has been provided, the PCR then serves an important financial function as coders and other finance staff will use what the EMS provider has documented to make decisions as to how the EMS services should be billed. Therefore, the PCR must accurately reflect the care given so the EMS provider accurately bills for the EMS services (accurate documentation = accurate billing = appropriate reimbursement). Illegible or incomplete documentation can cause errors in billing, resulting in the EMS provider receiving insufficient or inappropriate payment, or a denial.

The EMS Billing Process

Although EMS professionals should not be burdened with the inner financial workings of their EMS organization, a basic understanding of the billing process as it relates to PCR documentation and reimbursement is vital. In training programs, EMS professionals are often warned of the dangers of tunnel vision in patient care and told to "look at the big picture." This great clinical advice must also be our approach to documentation. So, let's take a look at the big picture. What happens in the billing department of the average EMS organization after the PCRs are turned in at the end of the shift?

First, the PCR is reviewed by a billing specialist and compared to essential dispatch information. In this examination, a few initial things should be assessed: Is it complete? Is it legible? What was done on the call? This review will take only a few minutes under the trained eye of the billing specialist to gain a basic understanding of the care that was given.

Next, a diagnosis code is assigned to the chief complaint documented on the PCR. Medical billing always depends upon a diagnosis. Health care providers use the International Classification of Diseases, or ICD9, system for attaching a code to a diagnosis. The **ICD9** system enables coders to communicate specific information about the diagnosis from thousands of ICD9 codes covering every imaginable diagnosis or condition. EMS providers also used the ICD9 system until the implementation of **condition codes** by Medicare in 2005. Condition codes reduce the ICD9 codes into generalized categories based upon the patient's condition. Following a condition code, a **procedure code**, or **CPT/HCPCS** (HCFA Common Procedure Coding System), assigns a level of service to the claim. Whereas the ICD9 or condition code designates the condition of the patient, the HCPCS procedure code communicates the complexity of care the patient required (see Table 3-2). Based upon the level of licensure of the EMS organization and responding personnel, the dispatch information, and the patient's condition ascertained through assessment and EMS treatment, one of these 11 procedure codes will be assigned to the claim. The level of service determines reimbursement according to the Medicare fee schedule for Medicare patients.

During this process the PCR is transformed from a clinical document into a financial document, becoming a "claim" that is billed to Medicare, Medicaid, and/or private insurance companies (or directly to the patient). Claims are usually billed electronically using universal data fields for placement of key information from the PCR. Because Medicare is the

Table 3-2	EMS Procedure Codes
A0425	Ground mileage, per statute mile
A0426	Ambulance service, advanced life support, nonemergency transport, level 1 (ALS1)
A0427	Ambulance service, advanced life support, emergency transport, level 1 (ALS1-Emergency)
A0428	Ambulance service, basic life support, nonemergency transport (BLS)
A0429	Ambulance service, basic life support, emergency transport (BLS-Emergency)
A0430	Ambulance service, conventional air services, transport, one way (fixed wing) (FW)
A0431	Ambulance service, conventional air services, transport, one way (rotary wing) (RW)
A0433	Advanced life support, level 2 (ALS2)
A0434	Specialty care transport (SCT)
A0435	Fixed-wing air mileage, per statute mile
A0436	Rotary-wing air mileage, per statute mile

largest payer source to EMS, we will use it as an example of the process as the claim leaves the EMS organization and arrives electronically at the Medicare carrier.

At Medicare, the claim is electronically reviewed and will follow two possible paths to payment or denial. First, if the procedure code has not been flagged for medical review and all components of the claim are technically correct, the claim will be set up for electronic payment. Second, if the procedure has been deemed by the carrier to require review prior to payment, the Medicare carrier will send a letter to the EMS organization requesting a copy of the PCR. Once the PCR has been received, a medical review nurse will conduct a formal medical review to evaluate the claim for medical necessity. Once a decision has been made, payment is made to the EMS organization or a denial letter is sent to the provider stating the reason for the denial. If Medicare denies a claim for EMS services, the EMS provider does have appeal rights and many denied ambulance claims are ultimately paid in the appeals process; but the process is time consuming and wastes valuable resources. In the appeals process, EMS providers are often required to obtain additional information from physicians and hospitals, which is usually difficult, if not impossible. Clearly, the time to document all pertinent information pertaining to the patient's history, condition, and treatment given is at the time of the event.

Medicare Basics

President Harry S Truman's vision to provide a national health insurance plan led to the creation of government health care coverage for the nation's elderly under Social Security. On July 30, 1965, President Lyndon Johnson (with former President Truman seated beside him) signed into law legislation creating the Medicare and Medicaid programs, undoubtedly the crown jewel of Johnson's Great Society (see Figure 3-1). Medicare's purpose is to provide basic health insurance coverage for persons age 65 and older and to individuals under age 65 who are disabled or are suffering from end-stage renal disease. Whereas Medicare is a federal program for the elderly and disabled, Medicaid is a state-administered program for those without income.

Medicare provides two primary types of benefits to beneficiaries known as Part A and Part B benefits. Most hospital-based EMS providers primarily bill Medicare Part A for their services, and non-hospital-based providers bill Medicare Part B. Medicare Part B is the primary payer for EMS. See Table 3-3 for examples of Part A and Part B benefits. Medicare is the trendsetter for all other insurance payers. The standards and guidelines developed and implemented by Medicare will almost always be adopted by other insurance companies. If Medicare pays for services in a certain manner, Medicare policy becomes the standard that will dictate payment practices for other payers.

Medicare is not the government's insurance company. Although it is often referred to as the "Medicare program," Medicare is fundamentally a group of laws or statutes. Understanding that Medicare is law will affect your approach to PCR documentation. Congress delegates oversight of Medicare to the Centers for Medicare and Medicaid Services (CMS). CMS in turn

Table 3-3 Medicare Part A versus Medicare Part B	
Medicare Part A	**Medicare Part B**
Hospital services (inpatient)	Physician services
Skilled nursing services	Outpatient hospital services
Hospice and home health care	Lab, physical, and occupational therapy
Most people are covered at age 65 without charge	Optional: Requires enrollment and has a monthly premium for coverage

delegates Medicare's day-to-day operation to private companies that have performance-based contracts with CMS to process, pay, or deny claims submitted to Medicare.

Medical Necessity

Medicare provides payment for basic health care. Medicare must be selective and pay only for medically necessary services in order to have enough money to pay for the health care of those presently enrolled in the program and those who will receive benefits at the time of retirement or disability.

Medicare recipients, known as beneficiaries, receive benefits as a result of the Medicare deductions that have been taken out of their paychecks during their years of employment. Although beneficiaries have earned their Medicare benefits and are entitled to them, these benefits are not automatic. The health care service must meet the medical necessity requirements. Health care providers, in a sense, earn the benefits for the beneficiary by medically necessary services. This is the reason for documentation-based health care. EMS providers not only record the clinical care given but also accurately document information related to the medical necessity of that care. Medical necessity and documentation are inseparable. (Medical necessity will be discussed in depth in Chapter 8.)

Medical Review

Medical review asks two simple questions:

- Does the EMS service meet coverage requirements?
- Does the documentation support the medical necessity for the EMS services?

The goal of medical review is to reduce Medicare payment errors by identifying billing errors made by health care providers. You can't pay for everything, neither can Medicare. Medical review is done by many dedicated professionals charged with upholding the directives set forth by Congress and the CMS. They take their jobs very seriously, and like you, they operate under written guidelines outlining the exact conditions for coverage and the required documentation in order for the service to be considered medically necessary. Like you, they have a quality process that evaluates their medical review decisions for accuracy based upon written standards. Medical review staff want to make the correct payment decision for the EMS provider and the Medicare beneficiary, and they rely on the PCR documentation to do so. They ask, Is the PCR legible, accurate, and complete? Does it capture the full picture of the EMS event by accurately reflecting the treatment and transportation provided for the patient? Does the PCR reflect that the medical care and the transport were medically necessary?

Medicare Denials

Medicare denials are a serious problem for EMS providers and our patients, costing the industry millions of dollars each year in lost revenue and resources. When claims for EMS services are denied:

- EMS providers are denied the revenue needed to maintain EMS systems.
- Beneficiaries are denied their Medicare benefits and are possibly held responsible to pay the EMS bills themselves.

Many beneficiaries do not understand why Medicare determined their ambulance trip to be "unnecessary," and the next time EMS services are needed they might think twice before dialing 911. So, when we document well, we are also assisting our patients in obtaining the Medicare and insurance benefits that are rightfully theirs. Provide your patients with the entire package of excellence: give them excellent patient care, and then document the care accurately with excellence.

Summary: Return to Case Study

ON TARGET

EMS professionals never make response or treatment decisions based upon reimbursement. Patient care decisions are always made on what is best for the patient. Money must have no part in it. Documentation must never be exaggerated or embellished in order to encourage payment.

The finance department receives the Explanation of Benefits (EOB) within a week, stating that the claim for your call was denied, as "documentation does not support medically necessary." As your company's billing specialist reviews the EOB, she comments: "What's up with Medicare? I guess they want to get up to their quota on denials. Oh, well, the patient's family will have to pay this one." A few months later a bill for $1,500 arrives at the home of the patient's family with an explanation attached that Medicare denied their claim. With the life insurance money all but depleted, family members wonder how they will ever pay the bill. Now more than ever, they can't help but feel the sting of pain each time they see an ambulance.

The EMS provider has not only professional responsibilities relating to clinical care but also financial responsibilities in accurately documenting the patient care so accurate billing and payment decisions can be made. The responsibilities of the EMS professional do not end at the end of the stethoscope but at the end of the pen after the care has been accurately documented.

After the patient has arrived at the destination facility, the PCR is completed. The clinical tasks end and the financial tasks begin as the PCR is completed with integrity. EMS professionals never make response or treatment decisions based upon reimbursement. Patient care decisions are always made on what is best for the patient. Money must have no part in it. Documentation must never be exaggerated or embellished in order to encourage payment.

CHAPTER REVIEW

Review Questions

Please refer to Answers to Chapter Review Questions at the back of this book.

1. Describe the link between documentation and reimbursement.

2. Describe the importance of the EMS professional having a basic understanding of Medicare history and process.

3. Describe how medical necessity is established.

4. Evaluate the PCR in Figure 3-2 and answer the following questions.
   - Based upon this PCR, how would you evaluate this documentation?

   - Based upon this PCR, how would you evaluate the patient care?

- Based upon this PCR, would it be easy for billing staff to bill for the EMS services?

- Place yourself at the desk of a Medicare Utilization Review nurse. Would you pay or deny the claim for these EMS services? Why?

EMS Documentation PCR Example							

Incident Location: *1425 W. Court*   City: *Smallville*
Report Number: *2005001425*   Miles: Start: *00* End: *20*
Incident Date: *5/25/05*   Medical Control: *none*
Sending Facility: Receiving Facility: *SJMC*

Last Name *Jones*	First Name	*Jane*	Middle *E*

Mailing/Home Address: *1425 W. Curt*

City: *Smallville*	County: *Iowa*	State: *OK*	Zip: *70000*

SSN: *100-00-1000*

Age: *70*	Date of Birth: : *4/12/*

Phone: *(918) 400-0000*

Billing Information:
Name/Company:
Insurance Company Address:
Group/ID Number: *100-00-10000B*   Medicare #:
Self-Pay: *No*

Chief Complaint: *GI Bleed*
Past Medical History: *GI Bleed*

Current Medications: *Prilosec, NTG, Premarin, HCTZ*

Allergies: *none*

Patient Found: *supine*

Stretcher Necessary? *yes*   Reason: *GI Bleed*

TIME	PULSE	RESP	BP	LOC	SAT	EKG
1600	*100*	*16*	*100/60*	*X4*	*95%*	*SR*
1615	*100*	*16*	*110/60*	*X4*	*95%*	*ST*
1630	*100*	*16*	*120/70*	*X4*	*95%*	*ST*

TREATMENT: RESPONSE:

1. *O2 15L NRB*   *Sat*
2. *IV NS W/O*   *BP*
3. *Monitor*
4.

Narrative:

Found patient supine in bed @ residence. C/O abdominal pain X 1 day. Vomited black emesis X 3 today. Transport to SJMC in poc. (unreadable signature)

Crew 1: *(Unreadable signature)*   Crew 2: *Bob*
Smith *EMT*

FIGURE 3-2
PCR Example

5. Evaluate the PCR in Figure 3-3 and answer the following questions.

EMS Documentation PCR Example

Incident Location: *2405 Broadway*   City: *Commonville*
Report Number: *200600017777*   Miles: Start *00* End: *16*
Incident Date: *5/15/2006*   Medical Control: *By Protocol*
Sending Facility: *Emergency Call*   Receiving Facility: *University*

Last Name *Deux*   First Name *Betty*   Middle *Ann*
Mailing/Home Address: *2405 Broadway*

City:	County:	State:	Zip:
Commonville	*Smith*	*Iowa*	*55555*

SSN: *000-00-0001*

Age: *70*   Date of Birth: *3/15/1936*

Phone: *(641)500-0000*

Billing Information:
Name/Company: *Blue Cross Supplement*
Insurance Company Address: *P.O. Box A, Denver, CO*
Group/ID Number: *92000*   *000-00-0001B*
Medicare #: *000-00-0001B*   Medicaid #: *Not Applicable*
Self-Pay: *No*

Chief Complaint: *Chest Pain w/Arrhythmia (Ventricular Tachycardia)*
Past Medical History: *Cardiac (Myocardial Infarction in 1999, CABG in 2000)*

Current Medications: *Toprol, Baby Aspirin*

Allergies: *Iodine*

Patient Found: *Seated in recliner, hand over chest w/respiratory distress*

Stretcher Necessary? *Yes*   Reason: *Life/Threatening Arrhythmia*

TIME	PULSE	RESP	BP	LOC	SAT	EKG
0600	180	24	60/40	*Person/Place/Event Only*	70%	*Ventricular Tachycardia*
0610	100	20	70/48	*Alert/Oriented*	90%	*Normal Sinus*
0615	98	16	110/78	*Alert/Oriented*	95%	*Normal Sinus*

TREATMENT:	RESPONSE:
1. *EKG monitor*	*Ventricular Tachycardia noted*
2. *Immediate Cardioversion @ 100 Joules*	*Conversion to Sinus Rhythm*
3. *O2 15 Non-Rebreather Mask*	*Increased Saturation*
4. *15 NS w/18 gauge catheter in Left Brachial*	*No pain/infiltration or redness*
5. *Amiodarone 150/mg IV*	*No further Arrhythmia*

Narrative: Substernal chest pain (began 30 minutes prior to arrival) radiating to left arm w/respiratory distress. Onset at rest. Pt states, "I feel a flutter in my chest." Assessment: General: Pale/diaphoretic, w/severe chest pain (9:10) HEEJ – No JVD. Respiratory: Symmetrical respirations. Rate and rhythm regular. Decreased expansion, clear bilateral breath sounds. CV: Neck veins flat. Mild peripheral edema. Abdomen: Rounded/Soft w/no pain or tenderness.

Crew 1: *A. Smith, Paramedic*   Crew 2: *E. Jones, AEMT*

FIGURE 3-3
PCR Example

- Based upon this PCR, how would you evaluate this documentation?

- Based upon this PCR, how would you evaluate the patient care?

- Based upon this PCR, would it be easy for billing staff to bill for the EMS services?

- Place yourself at the desk of a Medicare Utilization Review nurse. Would you pay or deny the claim for these EMS services? Why?

6. Evaluate the PCR in Figure 3-4 and answer the following questions.

EMS Documentation PCR Example			
Incident Location: *Regional Medical Center*		City: *Centerville*	
Report Number: 20060002005		Miles: Start: *00* End: *25*	
Incident Date: 7/1/2006		Medical Control: *None*	
Sending Facility: *Mercy ED*		Receiving Facility: *VAMC*	
Last Name *Smith*	First Name *John*		Middle *D.*
Mailing/Home Address: *Rt 1 Box 23*			
City: *Centerville*	County: *Small*	State: *Minnesota*	Zip: *55555*
SSN: *000-00-0001*			
Age: *85*	Date of Birth: *9/15/1921*		
Phone: *(763) 555-5555*			
Billing Information: Name/Company: *None* Insurance Company Address: Group/ID Number: Medicare #: *000-00-0001B* Self-Pay:		Medicaid #: *00001555-5*	
Chief Complaint: *Hip Fracture*		Past Medical History: *Cardiac*	
Current Medications: *Insulin*			
Allergies: *PCN*			
Patient Found: *ED Room 5*			
Stretcher Necessary? *Yes*		Reason: *Unable to Walk*	

TIME	PULSE	RESP	BP	LOC	SAT	EKG
1900	88	16	132/88	*Alert and Oriented*	95%	*Not Applicable*
1930	80	16	130/80	*Alert/Oriented*	95%	*Not Applicable*

TREATMENT:	RESPONSE:
1. *O2 @ 4L per Nasal Cannula*	

Narrative: *Transfer patient to VAMC for treatment of left hip fracture.*

Crew 1: *R. Johnson, Paramedic*	Crew 2: *K.Wood* EMT

FIGURE 3-4
PCR Example

- Based upon this PCR, how would you evaluate this documentation?

- Based upon this PCR, how would you evaluate the patient care?

- Based upon this PCR, would it be easy for billing staff to bill for the EMS services?

- Place yourself at the desk of a Medicare Utilization Review nurse. Would you pay or deny the claim for these EMS services? Why?

7. Evaluate the PCR in Figure 3-5 and answer the following questions.

EMS Documentation PCR Example	
Incident Location: *Community Hospital*	City: *Communityville*
Report Number: *200600015*	Miles: Start: *00* End: *5*
Incident Date: *1/15/2006*	Medical Control: *By Protocol*
Sending Facility: *Emergency Call*	Receiving Facility: *Golden Manor N.H.*

Last Name *Smith*	First Name *Irene*	Middle *B.*

Mailing/Home Address: *Golden Manor Nursing Home P.O. Box 100*

City: *Communityville*	County: *Benton*	State: *MN*	Zip: *55555*

SSN: *000-00-0001*

Age: *64*	Date of Birth: *1/2/1941*

Phone: *(952)555-1111*

Billing Information:
Name/Company:
Insurance Company Address:
Group/ID Number:
Medicare #: *000-00-0001B*   Medicaid #:
Self-Pay:

Chief Complaint: *G Tube Replacement*	Past Medical History: *CVA*

Current Medications: *Unknown*

Allergies: *Unknown*

Patient Found: *Community Hospital 4-11*

Stretcher Necessary? *Yes*   Reason: *Return to N.H. post G Tube Replacement*

TIME	PULSE	RESP	BP	LOC	SAT	EKG
1730	88	16	140/P	X2		

TREATMENT:	RESPONSE:
Transfer only	

Narrative:
Transfer to Nursing Home after G Tube Replacement.

Crew 1: *N. Smythe*	Crew 2: *E Tanner*

FIGURE 3-5
PCR Example

- Based upon this PCR, how would you evaluate this documentation?

- Based upon this PCR, how would you evaluate the patient care?

- Based upon this PCR, would it be easy for billing staff to bill for the EMS services?

- Place yourself at the desk of a Medicare Utilization Review nurse. Would you pay or deny the claim for these EMS services? Why?

Critical Thinking

Please refer to Answers to Critical Thinking Discussion Exercises at the back of the book.

1. Reimbursement decisions are often made by medical review nurses. Is this a problem for EMS, and what impact does this have on PCR documentation?

2. In thinking through the process of a PCR being evaluated for medical necessity, does documentation of patient care ever need to educate?

3. What value would there be in EMS professionals spending time in a billing department? Likewise, would there be value in billing staff riding with field crews or would this lead to a conflict of interest?

Action Plan

1. Find out who the Medicare carrier is in your state and visit its website. Conduct a search of its website on ambulance policy and note anything pertaining to "documentation requirements."

2. If appropriate in your EMS organization, schedule an appointment to visit the billing department. Spend an hour or two learning about your organization's billing process.

3. Many EMS organizations have Medicare Compliance Programs. If your company has such a program, become familiar with it.

Practice Exercises

1. Evaluate the narrative statement in Figure 3-6a. What problem(s) do you identify with this statement? Rewrite this statement to be more effective (Figure 3-6b).

Narrative Snapshot
Vital signs not taken – short distance to nursing home.

FIGURE 3-6a
Narrative Example

Narrative Snapshot

FIGURE 3-6b
Narrative Example

2. Evaluate the narrative statement in Figure 3-7a. What problem(s) do you identify with this statement? Rewrite this statement to be more effective (Figure 3-7b).

Narrative Snapshot

Patient bed-confined per history.

FIGURE 3-7a
Narrative Example

Narrative Snapshot

FIGURE 3-7b
Narrative Example

3. Evaluate the narrative statement in Figure 3-8a. What problem(s) do you identify with this statement? Rewrite this statement to be more effective (Figure 3-8b).

Narrative Snapshot

Bed-confined patient transferred back to nursing home.

FIGURE 3-8a
Narrative Example

Narrative Snapshot

FIGURE 3-8b
Narrative Example

CHAPTER 4

Legal Responsibilities and EMS Documentation

Key Ideas

Upon completion of this chapter, you should know that:

- The EMS professional will be held to a higher standard of accountability as clinical capabilities, scope of practice, and complexity of interventions increase, making proficiency in documentation essential.

- EMS professionals are responsible to the public and to the government in their professional practice, and PCR documentation is a reflection of this.

- The Patient Care Report serves a number of important functions within the legal system.

- PCR documentation must support the EMS professional's compliance with scope of practice and standard of care.

- Negligence in PCR documentation can be a violation of the standard of care.

- Integrity in PCR documentation is a moral and ethical responsibility of every EMS professional.

FIGURE 4-1
(Courtesy Lakes Region EMS, North Branch, MN)

Even though it has been two years since the endotracheal tube incident, the sore spot in your professional psyche remains. You feel the stigma of doubt when you're around members of management, and you worry your peers think your care is inferior. To you, this is a great tragedy, because excellent skills and perfect scene choreography have been the hallmarks of your career. The endotracheal tube was correctly placed in the field but was displaced by the staff at the Emergency Department when they moved the patient. It was your documentation, not your clinical skills, that let you down. During many years of hard work to perfect your skills, you neglected to attain proficiency in documentation, which would have supported and protected your career. You endured the six months of quality review, but couldn't shake the stigma you felt, so you opted to take another position in a neighboring state in a final attempt to jump-start your career.

The family of the patient that died in the MVC, however, is still working through their grief. In an effort to gain closure on their loved one's death, they visit one of the nurses who had assisted in taking care of their loved one the day he died. The nurse explains the course of treatment the patient received while in the Emergency Department that day, assuring them the staff did everything possible. However, the nurse inadvertently makes reference to the breathing tube having been in the wrong place. The family, who were attempting to work through grief by seeking out office information, are instead sitting in an attorney's office three days later.

The attorney patiently and sympathetically listens to the family's account of their loved one's death. They relate to him what they heard from the hospital nurse about the breathing tube having been in the wrong place. Intrigued, the attorney promises to investigate and promptly orders his staff to get the ambulance documentation. The attorney assures the family that the Patient Care Report *will tell them everything*.

Two weeks later, the family becomes the plaintiff and you and your former employer become defendants in a $10-million wrongful death lawsuit. To be continued.

Questions

Please refer to Answers to Case Study Questions at the back of this book.

1. If your patient care were put to the test of a negligence lawsuit, do you think it would it affect your relationships with your peers?

2. What is the difference between documentation and patient care?

3. Is it more likely for someone who is angry or grieving to file a lawsuit? Why or why not, and how does this tie to documentation?

Introduction

Usually in talking about the legal aspects of documentation, we make reference to the two most common legal mottos in EMS:

- "CYA."
- "If you didn't write it down, you didn't do it."

These expressions, uttered from the mouths of many EMS veterans, reveal the tremendous lack of understanding regarding the legal responsibilities associated with documentation. They limit accountability in documentation to staying out of trouble.

The expression "CYA" is problematic for a number of reasons. First, it suggests the "bare minimum" is the goal in PCR documentation. You can be sure the bare minimum will leave your backside "bare" if you and your PCR end up on the witness stand. Second, it also intimates the primary purpose of documentation is to "cover ourselves." What are we trying to cover? Are we covering ourselves (and our profession) for what we should have done but failed to do? The statement "If you didn't write it down, you didn't do it" also reveals a very shortsighted approach to EMS documentation. Simply writing it down so that you can demonstrate it was done is not enough.

When staying clear of the plaintiff's attorney was the main objective, legal accountability in EMS was usually met simply by delivering quality patient care. Today, our legal responsibilities go well beyond the avoidance of lawsuits. This chapter takes a much different approach as we examine the legal responsibilities of the EMS professional and the relationship to PCR documentation.

KEY TERMS

Note: Page numbers indicate where the following key terms and definitions first appear.

tort (p. 48)	malpractice (p. 50)	standard of care (p. 52)
negligence (p. 50)	scope of practice (p. 52)	False Claims Act (p. 55)

The Legal Challenges of EMS

The legal challenges of EMS are far greater than in years past, when the paradigm of documentation centered only on covering yourself just enough to stay clear of lawsuits. Today's legal challenges are in maintaining accountability in legal relationships to the public and federal and state governments. EMS providers and professionals are responsible to the public to provide quality patient care and to the government to demonstrate compliance with laws and regulations. These challenges impact patient care *and* documentation. In our litigation-happy society, EMS providers can be prime targets for civil lawsuits.

On a positive note, television has been a great educator. Programs such *Rescue 911* in the 1980s and early 1990s continued educating the public on EMS. A typical thrilling episode almost always began with a caller dialing 911. The 911 communications center quickly dispatched EMS resources and provided pre-arrival instructions to the caller, enabling the patient care to begin prior to EMS arrival. Once on scene, EMS personnel quickly stabilized, provided treatment to, and transported the patient with perfect results, week after week. *Rescue 911* was great for EMS and informed the public better than the best public relations campaign.

On the flip side, as *Rescue 911* educated, it also raised expectations. Today's EMS professionals practice under the bright lights of public scrutiny. From the stage of busy intersections to living rooms, patient care is delivered in front of bystanders and family members. The public expects not only pre-arrival instructions when they dial 911 but also the highest level of EMS care as they are treated and transported to a medical facility. Because the public also expects favorable outcomes, when results are less than favorable, patients and families hold EMS systems to the same level of accountability as physicians and medical facilities. A poor outcome must be someone's fault, and therefore they have a "right to the compensation that is due them." Accurate and complete PCR documentation is very often the only difference between success and failure in litigation. A poorly documented Patient Care Report hands the case to the opponent on a silver platter.

Accountability to federal and state regulations is an expectation of all health care providers, including EMS organizations. EMS organizations and professionals must comply with federal health care legislation (specifically HIPAA and EMTALA), federal health care programs (such as Medicare and Medicaid), and state regulatory oversight of EMS practice. Compliance holds EMS providers responsible to practice according to the law and is placing greater demands on PCR documentation. As the mirror of professional practice, PCR documentation reflects our relationship to the legal standards and regulations we practice under.

Public Accountability—Civil and Administrative Law

The EMS professional is also held accountable to the public to practice within legal boundaries that have been set by law.

Legal Structure

A basic knowledge of the structure and function of the legal system will enable the EMS provider to understand how PCR documentation operates within legal processes. Our legal system is built upon four types of law. Constitutional law, based upon the United States Constitution, establishes the structure and function of our government and legal system. Legislative law, established by the legislative branches of federal and state governments, produces the statutes or laws that govern EMS practice. Common law seeks to apply the same processes and principles that were used to arrive at decisions in past legal cases to present legal cases. Administrative law deals with regulations set forth by federal and state agencies, such as state EMS agencies, adding detail to statutory law in the form of regulations.

There are two types of legal action: criminal and civil. Criminal action is in response to violations of laws that are meant to protect society. A criminal conviction results in punishment such as a prison sentence or a monetary fine. Civil action is in response to a noncriminal conflict between two or more parties and results in monetary damages awarded to the winning party. A **tort** is a type of civil action that seeks to make amends for wrongs committed against another person, such as medical malpractice. Both criminal and civil laws govern the practice of EMS professionals.

tort
A type of civil law that seeks to provide monetary relief for harm or damages, which are usually a result of negligence.

Legal Process

A basic knowledge of the legal process will also assist EMS personnel in understanding the critical role of PCR documentation. The following example demonstrates both how the PCR is central to the legal process and how the PCR and the patient care itself are synonymous.

Imagine you have been sued. Despite your best efforts, the attempted resuscitation of a cardiac arrest patient resulted in an unfavorable outcome due to an allergy to an anti-arrhythmic medication. The family's unmet expectations, fueled by grief and anger, have led them to retain an attorney. After review of your Patient Care Report, dispatch tapes, state EMS laws, and local protocols, the attorney advises his client, the *plaintiffs*, that he believes their case has merit. He files a lawsuit naming you, your partner, your company, and your medical director as *defendants*. The *complaint* alleges that you were negligent because you failed to provide care in line with local protocols, or standard of care.

After you and the other defendants have been served the complaint, your attorneys issue a written response, called the *answer* to the claim(s) made in it. Your company's attorney notifies you will be interviewed during the *discovery* phase as they prepare to defend you and the other defendants. Six months after the call you are called for a *deposition* by the plaintiff's attorneys. At the deposition you are placed under oath, questioned in great detail regarding the care you gave the patient, and cross-examined by the plaintiff's attorney. Rattled by the experience, you dread the day you will be placed on the stand and cross-examined again in front of a jury.

Two long, drawn-out years after the EMS call, the case finally goes to court. By this time you barely remember many of the details of the call, let alone how you treated the patient. As the trial opens, the plaintiffs open their case, alleging the care you provided was negligent because you failed to follow the state's standard of care and local protocols established by your medical director. As the plaintiff's attorney begins his case, he unveils a six- by six-foot poster of your PCR placed directly in front of the jury. After his opening statements the attorney calls an expert witness to the stand, a nationally recognized EMS instructor. He meticulously walks the jury through your PCR line by line, pointing out the errors and omissions in your documentation, which substantiates the plaintiff's assertion that your care was outside the standard of care as defined by local protocols and therefore you were negligent.

It's all you can do to sit still. "This guy wasn't on the call with me. He has no idea how I took care of the patient. I did nothing wrong, and I certainly wasn't negligent." Two days later, you have the opportunity to tell your story when called to the stand as the primary defense witness. All eyes are upon you. "Now I remember them," you think, as you note the patient's family seated at the table directly in front of you. You're nervous, but while you've dreaded this moment, in a sense you're looking forward to it. It's time, finally, to give your side.

A lump begins to form in your throat as the attorney approaches you, laser pointer in hand. You realize this experience is not what you thought it would be and you're not going to be telling the story. The billboard-sized poster of your PCR will be doing the talking. The attorney begins by asking you to describe your training, licensure, and certification but quickly turns his attention, and laser pointer, to the PCR. Beginning with the date and incident number at the top of the PCR, the attorney goes through your documentation line by line. You regret not having taken more time to be clear in the care you provided, and not being able to read back, when asked, what you documented certainly hurts your credibility.

Long before the end of the second day on the stand, you realize you will never forget this experience. Almost every question begins with "you documented" and almost every answer is followed by "but you documented." It is constant and relentless. The attorney even asks you about the "ambu" bag you documented using: "What does 'AMBU' stand for?"

Perhaps the worst comes when you are asked: "How can you say you verified the medication dose when you didn't document that you verified it? Nowhere in the record did you document that you verified the medication dose, nor did you document that the patient had allergies. Does it not say in this EMS textbook that you are to verify the 'Five Rights of Medication Administration'? Please explain to the court the Five Rights of Medication Administration." At the end of the trial, the jury *deliberated* for six hours before returning a *verdict* for the plaintiffs, awarding them several million dollars. The star witness on the stand was really your Patient Care Report and, sadly, it let you down.

If your care is ever the subject of legal action, it will certainly be a life-changing event. EMS professionals who have gone through this experience attest to its being a very stressful process that lingers for several years and often changes the course of, or even ends, an EMS career.

Civil Law—Negligence and Malpractice

negligence
The failure, resulting in harm or injury, to perform an expected duty.

In health care the golden rule is to do no harm. When patient outcomes are poor and expectations have not been met, or the delivery of care has been inconsistent with standards of care, a claim of negligence in civil court can be brought against the EMS provider. **Negligence** is defined as professional conduct that fails to meet the established legal standards of care. Every EMS provider is required to provide care that is within the parameters of his or her licensure reflecting the training received. A major measuring standard for proving negligence is this: was the care and conduct of the EMS professional the same as what would have been given by another reasonably prudent EMS provider?

To establish negligence, the plaintiff must prove:

1. There was a duty to act according to a standard of care.
2. There was a failure to perform according to a standard of care.
3. The negligent act directly caused harm.
4. The negligent act also caused an actual loss.

malpractice
Negligence committed by a professional.

The legal principle of duty to act assumes, by reason of your license, that when on duty you will fulfill your duty to provide emergency services when called upon. Duty to act also assumes that you will continue to act until the proper transfer of care occurs. The failure to perform your professional duties is known as "breach of duty." In order for a lawsuit alleging negligence to be successful, all four elements listed above must be proven. Some states stipulate that EMS providers are immune from lawsuits, which is often perceived as a legal "get out of jail free card." However, many of these states allow for legal action against EMS providers when the performance of the EMS provider is regarded as "gross negligence" or "willful and wanton misconduct." **Malpractice** is slightly different from negligence in that it applies only to members of a profession. Malpractice is a type of negligence that implies a failure to use professional knowledge and skill and is therefore a failure to perform a professional duty.

The proof and defense of negligence rests upon the quality and accuracy of documentation. The PCR will be used to reveal whether or not the EMS professional's care was consistent with the standard of care and to prove whether or not care was negligent and brought harm to the patient. What is often perceived as negligence may simply be a documentation problem instead of an actual patient care problem. The EMS provider in the case study is the perfect example. Although the care was textbook perfect, the failure to document this care properly landed this EMS provider in the lap of a malpractice lawsuit. Consider the case of *Browning v. West Calcasieu Cameron Hospital.*

CASE LAW: *Billy A. Browning et al. v. West Calcasieu Cameron Hospital*[1]
The plaintiff, Billy A. Browning, brought a claim of negligence against a hospital-based ambulance service, alleging that EMS personnel were negligent in the care of his wife, Mrs. Jewell Browning. EMS personnel had been summoned to the Browning residence and evaluated Mrs. Browning for syncopal episodes with nausea and vomiting. PCR documentation stated the patient "felt fine and refused transport to hospital for evaluation." EMS personnel were subsequently recalled to the Browning residence within an hour of the refusal after the patient experienced cardiac arrest (and later expired).

The Browning case centers on allegations that EMS personnel violated standard of care. The plaintiff alleged that West Calcasieu Cameron Hospital (WCCH) was liable for the action of its EMS personnel in failing to (1) obtain and document Mrs. Browning's refusal according to WCCH's protocols; (2) warn her of the potential seriousness of her medical condition; (3) place her on a cardiac monitor on the first run; (4) transport her to a hospital on the first run; and (5) assess her medical condition properly and completely; and for (6) any other acts of negligence proven at a trial on the merits. In its ruling the court stated: "WCCH's protocols

[1]*Billy A. Browning et al. v. West Calcasieu Cameron Hospital,* 03-332 (La. App. 3 Cir, 2003).

give their EMTs certain guidelines to follow when procuring a patient's refusal. These protocols clearly provide the applicable standard of care. . . ."

Although this case centers on a patient refusal, the message is clear. Not only must patient care be delivered according to the standard of care, documentation must also support that care was delivered according to the standard of care.

Negligent Documentation

Documentation is a standard of care in EMS because:

1. A PCR is required by state regulations in many states. The state of New York mandates,

   a prehospital care report shall be completed for each patient treated when acting as part of an organized prehospital emergency medical service, and a copy shall be provided to the hospital receiving the patient and to the authorized agent of the department for use in the State's quality assurance program.[2]

2. Instruction in patient care documentation is established curriculum in the different levels of EMS training programs.
3. State and local protocols require documentation for every EMS patient encounter.
4. Most EMS organizations have policies and procedures in place that address requirements for documentation.

Documentation that fails to reflect care and treatment appropriately is a violation of standard of care. An EMS provider in violation of the standard of care can be found by the courts to be negligent. Because proper documentation on the PCR is now the standard, failure to document can be considered negligence. In the past, court cases against EMS providers have centered totally around the patient care as recorded on the PCR. Courts evaluated emergency care based upon the record of events and treatment using the documented care to compare the EMS provider's care to written regulations and care standards in order to render their verdicts. The bar was raised in the case of *Franscesca C. V. De Tarquino v. Jersey City.*

CASE LAW: *Franscesca C. V. De Tarquino v. The City of Jersey City*[3]

EMS personnel were called to a police station to care for Julio De Tarquino, who was reported to have been in an altercation with police officers. Mr. De Tarquino was assessed, treated, and transported to a local hospital for evaluation of a possible head injury. After being evaluated by the Emergency Department staff at the hospital, he was released into the custody of the police department. Four days later Mr. De Tarquino died of an epidural hematoma. The family filed suit against Jersey City EMS, the specific EMS professionals, and other defendants for negligence and wrongful death.

The plaintiff's case centered on the failure of the EMS personnel to document properly that Mr. De Tarquino had experienced nausea and vomiting while in EMS care. The EMS provider completing the PCR documented, "-N/V." The defendants acknowledged this notations indicated "negative for nausea and vomiting." The Emergency Department physician asserted he would have conducted a more in-depth evaluation of the patient if he had been aware of the nausea and vomiting and would have ordered a CT scan, which would have revealed the emergent condition of the patient. In discovery, it was also found that a second PCR had been completed that had the "vomiting" box checked. Plaintiffs asserted this was done to cover up the failure to document the victim's nausea and vomiting properly.

This case has stunning implications for EMS documentation. The New Jersey court ruled that failure to document constituted negligence by itself, extending negligence to the

---

[2]State of New York, *Electronic PCR Data Submission* (2004). Retrieved March 30, 2007, from www.health.state.ny.us/nysdoh/ems/policy/04-05.htm.

[3]*Francesca C. V. De Tarquino v. The City of Jersey City*, 352 N.J. Super. 450, 800 A. 2d 255 (2002).

omission of key patient information on the PCR. Although the De Tarquino case was an isolated one, it is the first court decision to suggest that negligence in documentation can be held to the same legal standard as negligence in other patient care activities. Keeping in mind the definition of negligence, what makes documentation negligent? Documentation is negligent if the documented care fails to reflect the standard of care, as defined by state regulations, local protocols, policies, and procedures.

Avoiding negligence in documentation takes deliberate action.

1. Understand and embrace that documentation is the highest form of professional expression.
2. Understand that documentation is the standard of care. Failure to follow the standard of care leaves you legally vulnerable.
3. Be meticulous in documentation. Unclear, incomplete, or inaccurate documentation will leave you vulnerable to allegations of negligence. Documentation *is* patient care. If documentation is negligent, then the assumption could be made that the patient care was also negligent.
4. Hold yourself accountable to excellence in documentation. As the EMS profession advances, so does our level of professional accountability. If we fail to hold ourselves accountable, law firms, civil courts, and federal and state governments will do it for us.

Administrative Law

When you opened the envelope from the state EMS office and placed the new card in your wallet, something profound took place. By reason of passing the exams, your state conferred upon you the legal right, through licensure, to practice EMS at a certain level. The state is empowered by the laws that govern EMS written by its legislature to grant you a license. The ability to practice is a privilege, and licensure is your entrance into the realm of professional accountability.

In fact, EMS practice is governed by laws. A common misconception is that the medical director grants you the privilege to practice in EMS under his or her license. Legal responsibility is not deferred to your medical director. You do not practice under the medical director's licensure but under your own. Although medical direction ultimately retains the highest level of legal responsibility over patient care activities, you will be held responsible for *your* actions. EMS practice is governed by scope of practice and standard of care.

Scope of Practice

scope of practice
Defined boundaries for professional practice.

Scope of practice refers to the boundaries of practice and defines the skills and procedures granted by your license as an EMS provider. Scope of practice in your area, whether defined by the state, county, or city, sets the limits for your license. Operating inside the boundaries of scope of practice provides safety, whereas operating outside those boundaries puts you in a very vulnerable position. Many states have developed scope of practice documents for each level of EMS licensure, and those that have not usually define EMS practice boundaries within their state EMS statutes. PCR documentation of each patient encounter must reflect that all aspects of patient care was within scope of practice. Violations in scope of practice can result in disciplinary action from the state EMS agency issuing the license.

Standard of Care

standard of care
Detailed written guidelines or standards that direct patient care.

Standard of care refers to written guidelines that direct patient care. Standard of care:

- Establishes a tool of comparison by which care can be evaluated.
- Defines what a reasonably prudent EMS provider would have done in your situation.

Standard of care, while based somewhat on state laws (specifically scope of practice), is usually defined by written protocols, training curriculum, and standard operating procedures. Standard of care is a ruler by which your conduct and practice will be measured against that of your peers. Violations of standard of care can result in allegations of negligence, resulting in civil and disciplinary action. PCR documentation is the written legal record of how you met the standard of care. Consider the case of *Henslee v. Provena Hospitals*.

CASE LAW: *Henslee v. Provena Hospitals*[4]

Plaintiffs in *Henslee v. Provena Hospitals* contended that EMS personnel refused to allow a physician to intubate a patient in anaphylactic shock but were subsequently unable to manage the airway themselves. By failing to manage the airway appropriately, plaintiffs alleged that the EMS provider was in violation of protocol and therefore the standard of care. The defendants failed to convince the court that the actions of the EMS personnel amounted to "simple negligence." Instead the court ruled that the actions rose to the level of "gross negligence" because PCR documentation indicated the failure to follow local protocol. This case demonstrates the importance of PCR documentation being consistent with the standard of care.

Governmental Accountability

As EMS professionals, we are accountable not only to the public but also to federal and state governments. As noted previously, EMS professionals practice not only within the boundaries and measure of scope of practice and standard of care but also in compliance with federal law. This facet of EMS practice is usually miles away from the professional psyche of EMS professionals. Governmental accountability is expressed in ethical conduct in all aspects of patient care activities that is equally reflected in PCR documentation. This discussion examines federal legislation that impacts PCR documentation:

- Health Insurance Portability and Accountability Act of 1996 (HIPAA)
- Emergency Medical Treatment and Active Labor Act of 1986 (EMTALA)
- False Claims Act

Compliance with these federal regulations is dependent on the EMS professional:

- Having an understanding of one's legal responsibilities with respect to federal law.
- Possessing high ethical standards in EMS practice.
- Making a commitment to meeting one's solemn obligations to the federal government.

The EMS professional's compliance is expressed in accurate PCR documentation.

The PCR and HIPAA

The privacy and security of patients' health information is one of the foremost responsibilities of the EMS professional. The PCR does not belong to you; it is the patient's property. It contains the patient's private demographic and personal medical information, and upon arrival at the scene you become the legal custodian of this information. Although the HIPAA legislation is broad in scope, its Privacy Rule has immense importance for EMS providers. The Privacy Rule protects patients' medical records, allowing patients more control over their health information, and restricts the release of protected health information to only what is needed for the purpose it is to serve. The EMS professional's obligations to HIPAA are simple: the PCR is protected information under federal law. Violations of the HIPAA Privacy

[4]*Henslee v. Provena Hospitals*, 369 F. Supp. 2d 970 at 980 (2005).

FIGURE 4-2
PCR Documentation and HIPAA Compliance

Rule are taken very seriously by the federal government and can have grave consequences. Figure 4-2 provides guidelines for protecting patient's health information.

The PCR and EMTALA

Unit 11, Mercy Hospital. We are inbound Code Red with a Cardiac Arrest.

Mercy, Unit 11, we are diverting you to University. We're currently not accepting patients. We're on bypass until 2300 hours.

A fundamental right of all patients is emergency medical treatment. In years past, patients were often denied emergency treatment in hospitals as a result of ambulance divert policies. In response, Congress passed the Emergency Medical Treatment and Active Labor Act of 1986 (EMTALA). Known as the antidumping law, this legislation was designed to protect the rights of emergency patients. The intent of EMTALA is simple. When a patient seeks emergency treatment, the hospital must:

- Provide either examination or treatment for stabilization.
- Provide for appropriate transfer.
- Delay transfer until the patient has been appropriately stabilized.

EMTALA has serious implications, and Figure 4-3 lists the important documentation points for hospital diversions and emergency interfacility transfers. Remember your training: If the diversion of a patient to another facility or if an interfacility transport is inappropriate, contact your supervisor and/or medical control at once.

The PCR and the False Claims Act

Being truthful in documentation is the legal, moral, and ethical responsibility of every EMS professional. Our approach to PCR documentation must be as solemn as stepping up to the witness stand, raising your right hand, swearing to tell the truth, and then doing so. Signing your name at the bottom of the PCR says, "This is the absolute truth." Falsification of a PCR is a serious matter and the EMS professional must be diligent to avoid documenting anything that is less then 100 percent accurate. Examples of false documentation include:

- Documentation that is incorrect or misleading. For example, if vital signs were taken off the medical record of a sending facility, obtained by another agency, or obtained by Emergency Medical Responders but recorded as *your* first set of vital signs, documentation has been falsified.
- Documentation that inaccurately establishes medical necessity in order to obtain reimbursement. For example, if a patient is documented as "bed-confined" when he or she is ambulatory, documentation has been falsified.

Diversions

1. Document the hospital diversion thoroughly:
   - Diverting facility
   - Facility diverted to
   - Time of divert
   - Reason for divert
   - Person advising you of the divert
   - Medical control involvement (if applicable)
   - Supervisor notification
2. Inform and explain the divert process to the patient, and document the patient's response using his or her own words, if applicable.
3. If you questioned or verbalized disagreement over the change in receiving facility, document the conversation.
4. If the destination hospital is of greater distance than the diverting facility, document the reason for not transporting the patient to the closest appropriate medical facility.

Interfacility Transfers

If, in your professional judgment, you believe you are being called for a transport that violates EMTALA, in that the hospital has within its capabilities the ability to stabilize the patient properly prior to transport, you must document this thoroughly. Document:

- Name of the physician caring for the patient and ordering the transfer
- How and why the patient is unstable
- Reason for immediate transfer prior to stabilization
- History of present illness, paying close attention to the treatment given at the sending facility

FIGURE 4-3
PCR Documentation and EMTALA Compliance

- Documentation that misrepresents the actual patient care. It is "looking good on paper" when the actual patient care wasn't so good. For example, when protocols state that a medication is to be administered at specific time intervals, such as sublingual nitroglycerin at five-minute intervals, the documentation automatically reflects that time minute interval, regardless of the actual time in between doses. Unless you actually record or note the time that you administer a medication, documentation has been falsified.

Falsified documentation can lead to serious consequences. Consider the case of *Anonymous Infant v. ABC Ambulance Company*.

CASE LAW: *Anonymous Infant v. ABC Ambulance Company*[5]

EMS personnel responded to an infant that was having febrile seizures. The EMS providers had been ordered by medical control to administer Valium for the seizures but were unable to follow through with the order because they had forgotten to bring the keys to their drug cabinet. During preparation for the case, it was discovered that the EMS provider had allegedly falsified vital signs and medication administration times. The jury awarded the plaintiffs $10.2 million.

False Claims Act
A federal law designed to discourage fraud against the government by allowing an individual to file suit on behalf of the government against those suspected of fraudulent practice.

The pressure to perform, the lack of training in documentation, and the desire to look good on paper and stay clear of the "QA Police" lead many to "make it all come out right on paper." This very dangerous practice will place you and your company at risk for criminal prosecution.

Health care fraud devours 10 percent of the nation's health care expenditures. The federal **False Claims Act**, which dates back to Abraham Lincoln in 1863, was amended in 1986 to address the growing problem of fraud by seeking to create a partnership between the

[5]Breakstone, While & Gluck, *Negligent Paramedic Care—Failure to Treat Infant Seizures, Altered Medical Records*. Retrieved March 30, 2007, from www.bwglaw.com./cr_ambulance.html.

government and the public. The False Claims Act allows a person to file a lawsuit, in partnership with the government, against those who are allegedly committing fraud. In EMS, the False Claims Act is primarily concerned with correct billing and that billed services meet the medical necessity requirements.

Compliance with federal regulations is a shared responsibility of the entire EMS organization at all levels: administration, billing, and clinical. Do not be fooled; this is your responsibility. Never allow yourself to be pressured into documenting something that is false or inaccurate. Never exaggerate or embellish any aspect of patient care documentation. Simply document the truth.

Summary: Return to Case Study

After a three-week trial, the case was finally resolved. It proved to be a grueling and painful experience. The six- by six-foot billboard of your PCR stayed in front of the jury for almost a week. The woman on the jury who kept staring at you as if you were incompetent was almost more than you could bear. You wanted to shout out to the court, "I am really a great EMS provider. You don't understand. This Patient Care Report is not me!" However, in the end the jury found you negligent and awarded the plaintiffs a very large sum of money. After returning home you decide it's time for a career change and make the decision to go back to school.

Gone are the days when the legal responsibilities of EMS providers began and ended with "CYA" in order to stay out of court. Today, the public and the government hold EMS professionals accountable to the standards established by law. The next chapters will offer you the information and the tools to produce PCR documentation that accurately captures the essential documentation elements for each patient encounter. In doing so, you will meet your legal responsibilities and demonstrate accountability as an EMS professional.

CHAPTER REVIEW

Review Questions

Please refer to Answers to Chapter Review Questions at the back of this book.
1. List the four types of law and how they function within the legal system.

2. List four things a plaintiff must prove in order to establish negligence.

3. List and describe how standard of care is established.

4. Why is accurate PCR documentation a standard of care?

5. Describe the difference between scope of practice and standard of care.

6. You document "NKDA" (no known drug allergies) on a PCR, but you actually never asked the patient whether he or she was allergic to anything. Is this false documentation? Why or why not?

7. An EMT is found to have administered an IV medication that is prohibited in his or her state of practice. The EMT has violated:

    *   Standard of care

    *   Scope of practice

8. List the measures that can be used to guard a patient's protected health information in PCR documentation in order to comply with HIPAA regulations.

9. You have an emergency patient and are being diverted to another facility. List the essential documentation elements related to EMTALA compliance.

10. Describe the purpose of the False Claims Act and how it relates to PCR documentation.

Critical Thinking

Please refer to Answers to Critical Thinking Discussion Exercises at the back of this book.

1. "Negligence in documentation is negligence in patient care." Do you agree with this statement? Why or why not?

2. What are the ramifications when scope of practice and standard of care are ill-defined? How does this affect documentation?

Action Plan

1. Research your state's scope of practice laws. You have a responsibility to know and understand the scope of practice laws governing your practice. If you discover that your state does not have an actual written scope of practice for your level of licensure, determine how scope of practice is defined in your state.

2. Review your organization's HIPAA and EMTALA policies and procedures.

3. Using your organization's protocols, compare your PCR documentation to your organization's standard of care.

Practice Exercises

1. Considering the legal aspects of documentation, what are your observations of the narrative statement in Figure 4-4a?

Narrative Snapshot
Patient ambulated to stretcher post fall.

FIGURE 4-4a
Narrative Example

*   What are the potential problems with this statement?

*   What additional information would you have obtained?

*   Assuming your assessment validated the information, rewrite this statement (Figure 4-4b).

Narrative Snapshot

FIGURE 4-4b
Narrative Example

2. Considering the legal aspects of documentation, what are your observations of the narrative statement in Figure 4-5a?

Narrative Snapshot
Patient fell in between the stretcher and hospital bed while being transferred into bed.

FIGURE 4-5a
Narrative Example

- What are the potential problems with this statement?

- What additional information would you have obtained?

- Assuming your assessment validated the information, rewrite this statement (Figure 4-5b).

Narrative Snapshot

FIGURE 4-5b
Narrative Example

3. Considering the legal aspects of documentation, what are your observations of the narrative statement in Figure 4-6a?

Narrative Snapshot
1 amp of D50W given IV—infiltration or redness noted.

FIGURE 4-6a
Narrative Example

- What are the potential problems with this statement?

- What additional information would you have obtained?

- Assuming your assessment validated the information, rewrite this statement (Figure 4-6b).

Narrative Snapshot

FIGURE 4-6b
Narrative Example

4. *Role-Play Exercise*: Using the PCR from the case study in Figure 4-7, as a group (or one-on-one) "try" the case in court. Begin by choosing one person to play the role of the defendant and another to play the role of the plaintiff's attorney.

If you are the plaintiff's attorney:

- What weakness in documentation/patient care do you see? How would you "try" your case?

- What is missing in the PCR documentation?

- How would you use the PCR to support a claim of negligence?

- What would you attempt to find in discovery?

- What questions would you ask the defendant?

If you are the defense attorney:

- How would you defend your client?

- What resources would you use in your defense?

EMS Documentation PCR Example						

Incident Location: *Hwy 9 and County 15*
Report Number: *20060001736*
Incident Date: *2/25/2006*
Receiving Hospital: *County*

Last Name	*Doe*	First Name		*John*	Middle	

Mailing/Home Address: *unknown*

City:	County:	State:	Zip:
uto			

SSN: ....

Age: *approx 40s*	Date of Birth:	*UTO*

Phone:

Billing Information: *See Hospital Records*

Name/Company:

Insurance Company Address:	Group/ID Number:
Medicare #:	Medicaid #:
Self-Pay:	Miles:

Chief Complaint: *trauma*

Current Medications: *unknown*

Allergies: *unknown*

Patient Found: *in vehicle*

Stretcher Necessary?	*trauma code*	Reason:

TIME	PULSE	RESP	BP	LOC	SAT	EKG
scene	*80/40*	*rapid*	*120*	▼		*sinus*
en route	*80/40*	*rapid*	*120*			*sinus*

TREATMENT: RESPONSE:

1. *spinal immobilization*   *Emergency Medical Responders*
2. *O2/IV/Monitor*
3. *ET Tube*

Narrative:
Trauma. Followed Trauma/ACLS protocols. Care turned over to hospital staff.

Patient Signature: *unable*

Crew 1:	*P. Crew*	*T. Smythe*

FIGURE 4-7
PCR Example

CHAPTER 5

Documentation Fundamentals

Key Ideas

Upon completion of this chapter, you should know that:

- EMS documentation must perform three vital functions in order to be effective. It must inform, educate, and integrate.

- Effective documentation is clear, complete, correct, consistent, and concise.

- Clear PCR documentation is legible and understandable, leaving the reviewer with only one interpretation of the EMS event.

- Complete PCR documentation communicates all essential elements of the EMS professional's management of the EMS event.

- Accurate PCR documentation provides a factual and truthful account of the EMS professional's management of the EMS event.

- Errors and omissions in PCR documentation must be corrected appropriately using a consistent and systematic approach.

- Consistent PCR documentation unifies the demographic, financial, clinical, and transfer of care components of the EMS event.

- Concise PCR documentation records only the relevant facts of the EMS event.

FIGURE 5-1
(Courtesy EMSA, Tulsa, OK)

CASE Study

oncerned by the quality of documentation in your EMS organization, the administration and the medical director have formed a committee to address how best to improve the quality of PCR documentation. Over the course of several months, the director, quality manager, billing manager, supervisors, medical director, and three representatives from the field discuss the problem of documentation quality and, as with most committees, everyone comes to the table influenced by the "special interests" of his or her own department.

The billing manager is concerned about the $1-million increase in accounts receivable due to the rise in nonbillable ambulance claims as a result of incomplete documentation. The medical director and the quality manager, as a result of a review of trauma arrests, are concerned that the inaccuracies and inconsistencies in PCR documentation are affecting data collection and ongoing research studies. The supervisors express interest over the decrease in unit-hour productivity if crews are expected to take more time to complete PCRs. The field representatives are concerned they're being put on the spot, pointing out the absence of a consistent standard for defining quality and the lack of agreement among medical direction, administration, and billing with respect to PCR documentation.

Listening intently to the discussion, the director expresses her concern about "the silo effect," explaining that it appears each department within the agency is functioning separately in its own "tower of tunnel vision." Seeing PCR documentation as central to the interests of the entire organization, she mandates a Documentation Compliance program to improve quality. This program will begin with mandatory peer review of PCRs.

Questions

Please refer to Answers to Case Study Questions at the back of this book.
1. Do you think there is anything wrong with the organization's approach to the "documentation problem"? Why or why not?

2. What will be necessary for this initiative to be successful?

3. Do you think your perspective on EMS documentation would change if you were placed in a QI role?

Introduction

The case study reveals many of the issues associated with PCR documentation within EMS organizations and the confusion as to how to improve documentation quality. Improving quality begins with acquiring an understanding of the fundamentals of effective documentation. Effective PCR documentation:

- Represents the defining characteristics of the EMS professional.
- Captures the EMS care given to the patient so EMS services can be billed accurately.

- Reflects that the care provided by the EMS professional was both within the scope of practice and consistent with the standard of care.
- Educates the non-EMS reviewer so the care and services are understood.
- Integrates the EMS patient into the continuum of care at the receiving facility.

Whether the PCR is reviewed for clinical, legal, or financial purposes, it must communicate all relevant facts of the EMS encounter. Many EMS providers fail to master documentation and become frustrated. On your first day in EMT school, if you were required to place a trauma patient in a KED device in three minutes, you would have either failed or left the building in frustration.

This chapter focuses on gaining a firm grip on effective EMS documentation. We will use the same approach for mastering any EMS skill, by breaking down the basic tasks of completing the PCR using the Five C's of Clinical Documentation.

KEY TERMS

Note: Page numbers indicate where the following key terms and definitions first appear.

electronic Patient Care Report (ePCR) (p. 65)
credible (p. 65)
concise (p. 72)

The Five C's of Clinical Documentation

In order for the PCR to be effective, documentation must be clear, complete, correct, consistent, and concise.

EMS Documentation Basic #1—Clear

electronic Patient Care Report (ePCR):
The electronic Patient Care Report is a computer-based PCR that provides for enhanced data collection, quality management, and system integration.

Effective and professional documentation is clear and understandable. This means that, first and foremost, PCR documentation must be legible. EMS systems are quickly transitioning to **electronic Patient Care Reports (ePCRs)**. While the ePCR has many advantages based upon the capabilities of the particular software, these products all have one benefit in common: they answer the problem of legibility. Because handwritten Patient Care Reports are at the mercy of the individual writing them, legibility in documentation has been and remains a significant problem for the EMS industry.

Illegible clinical documentation is a legal minefield. First, if you are involved in legal action, and documentation is not legible, the defense options are few. Think about it: a sloppy and unreadable PCR communicates poorly about the care that the patient received. Second, judgments are made about you as a professional based upon the quality of your documentation. Imagine being on the witness stand attempting to decipher an unreadable Patient Care Report. The care may have been perfect, but with imperfect documentation your competency as an EMS provider will be judged. Third, if documentation is not legible your credibility will be questioned. Credibility as a medical professional is everything; without it you have nothing. To be **credible** is to be believable. You want the reviewer of your PCR to believe in your competency and credibility. It is not worth the risk of losing your credibility as an EMS professional due to illegible documentation. Every EMS professional whose organization uses a paper PCR requiring handwritten documentation must take the time to assess the legibility of his or her documentation (see Figure 5-2).

credible
Believable and trustworthy.

FIGURE 5-2
When Legibility Is a Challenge

ON TARGET Nothing will rob you of your credibility quicker than sloppy documentation.

PCR documentation must also have only one interpretation. Documentation that is subject to interpretation opens the door to misinterpretation. Consider the incorrect example in Figure 5-3. In this example, the following missing or unclear information could leave a reviewer with an incorrect interpretation of the intervention:

- No information is provided regarding the manner in which the patient was prepped for endotracheal intubation.
- Documenting "+ placement" is unclear and fails to establish that endotracheal tube placement was properly verified.
- Documenting "− BS present" is unclear. Does this mean no breath sounds present or no bowel sounds present?
- Documenting "Verified placement w/no—changes" is unclear, leaving the reviewer to question how placement was verified and the meaning of "—changes."
- Documenting "+ETCO$_2$" is unclear, leaving the reviewer possibly to assume the device reading indicated improper placement on the endotracheal tube.

Now consider a correct example of this narrative snapshot in Figure 5-4. It shows that PCR documentation must be clear to the reader, leaving only the impression the EMS professional intended.

Abbreviations

The use of abbreviations in medical documentation has a sharp double edge. On one hand, abbreviations are convenient and enable the caregiver to speed up the documentation process. On the other hand, their use may be a source of misinterpretation and may lead to patient harm. Further, when Patient Care Reports are reviewed for legal and reimbursement purposes, abbreviations may be a source of confusion.

Opinions and practices of EMS agencies vary when it comes to abbreviations, running the gamut from free rein to published lists of acceptable ones. When using

---

Narrative Snapshot

Patient intubated with 8.0 cuffed Tube + placement, - BS present. Verifed placement w/no-changes, +EJCO2.

FIGURE 5-3
What Is Your Interpretation? An Example of an Incorrect Narrative

Narrative Snapshot
Oral pharyngeal airway inserted using appropriate technique. Hyperventilated with BVM attached to 100% oxygen prior to intubation. Cardiac monitor and pulse oximetry noted (NSR/85%). Using a 4.0 straight blade, trachea intubated with 8.0 cuffed endotracheal tube (no stylet used). Vocal cords visualized as tube passed. Endotracheal tube placement verified with: Equal bilateraral breath sounds present with no breath sound present over epigastrium, noted mist in tube, and continuous ETCO2 detection indicated appropriate placement of ET tube (O2 saturation @ 95%). Cuff inflated with 8 cc of air, and tube secured using commercial device at 23 cm. Re-verified placement with no changes. Continuous ETCO2 monitoring with appropriate waveform at all times. Re-verified tube placement during transport and each time patient was moved.

FIGURE 5-4

Not Open to Interpretation. An Example of a Correct Narrative

abbreviations it is important to remember the purpose of PCR documentation. Although we document for our own internal use, keep in mind that the PCR is used by those outside EMS for research and reimbursement purposes. Think beyond EMS. Abbreviations do save time and effort, but what may be a convenience when completing the PCR may prove a great inconvenience later if documentation is misinterpreted. Consider the following examples of incorrect and correct uses of abbreviations in Figure 5-5. In addition, Figure 5-6 provides guidance in the use of abbreviations. Use abbreviations with caution, and the best advice is to use only those on an approved list that your organization endorses.

Narrative Snapshot		
Incorrect	Correct	Rationale
Assessment revealed no left BS.	*Assessment revealed absent bowel sounds in left upper and left lower quadrants.*	"No left BS" could be misinterpreted as "no left breath sounds." Avoid the use of abbreviations when documenting a negative assessment finding.
1610: patient given .5 mg MS IV	*1610: 0.5 mg of morphine sulfate given in 5 ml of sterile saline for injection per pediatric protocol.*	1610: ".5" represents an improper decimal due to absence of leading zero, which could be misinterpreted as 5 mg. Always use a zero before a decimal point.
1615: UTO respirations	*1615: Respirations 20, with 95% oxygen saturation.*	1615: The acronym UTO or "unable to obtain" must be avoided. Instead document the reason information was unobtainable.
2 μg administered	*2 micrograms administered.*	"μg" for micrograms can be misinterpreted as milligrams, especially when handwritten.
NTG X3	*Sublingual Nitroglycerin administered – 0.4 mg @ 1600, 1605, and 1610 with no changes in vital signs and decrease in pain status from 10:10 to 4:10.*	The use of X to denote a number of treatments fails to provide clear and sufficient information as to the specific treatment provided.

FIGURE 5-5

Abbreviation Examples

1. Utilize only agency-approved abbreviations.
2. If abbreviations are used, an abbreviation key should be provided on the PCR.
3. NEVER abbreviate the names of medication or units.
4. NEVER use abbreviations on patient refusal forms.
5. Avoid abbreviations in the diagnosis and Medical Necessity Statements.
6. NEVER abbreviate the names of sending and receiving medical facilities.

FIGURE 5-6
Principles for the Use of Abbreviations

EMS Jargon

EMS jargon must also be avoided in PCR documentation. Because "shop talk" is not understood by those outside EMS, it opens the door for misinterpretation and reflects poorly on the EMS professional. The following are a few examples:

- **UTO:** "Unable to obtain" is a documentation hazard. (It could also be interpreted as IDIOT—I didn't investigate or think.) Why were you not able to obtain the information? What else did you fail to obtain? Documentation is not complete until you have noted all the information. NEVER document "unable to obtain" because someone WILL have to get this information, sooner or later. If you were truly unable to obtain the information, document the information you were unable to obtain and the reason.
- **PUTS:** "Patient unable to sign" is unacceptable and should never be used. Instead, document that the patient was unable to sign and the reason.
- **Ten Codes:** This or any other type of signal designation should not be recorded in PCR documentation. Definitions for signals vary among geographical regions and will not be understood by those outside your EMS system or region.

EMS Documentation Basic #2—Complete

One of the most important aspects of professional documentation is that it can be understood by the layperson and trained medical professional alike. This is vitally important because 10 years after the fact, you may be asked to recount the details of your care as if it were yesterday.

Documentation must be complete because complete PCR documentation communicates all essential elements of the EMS professional's management of the EMS event. Incomplete documentation is an open door for wrong assumptions being made about the patient care. In the same manner that illegible documentation leads to the assumption that the care was sloppy, inaccurate conclusions can be made that incomplete documentation equals incomplete care. Whether PCR documentation is done on paper or electronically, it must be complete. Electronic Patient Care Reports, while electronically composed, still become paper PCRs at the end of the EMS event.

PCR documentation that is complete:

- Captures all pertinent aspects of the EMS event. Ensure that all appropriate boxes, blanks, or data entry fields are filled in. If a field is not applicable, write out "not applicable."
- Answers the questions who, what, when, where, and how in order to capture all pertinent aspects of the EMS event.
- Is balanced documentation. A PCR that has a very detailed narrative yet has fields left blank is unbalanced and incomplete, as is a PCR that has all appropriate data entry fields complete with a blank or insufficient narrative.

EMS Documentation Basic #3—Correct

Documentation must also be correct, representing a truthful and factual reporting of the EMS event as it happened. It is essential for the PCR to capture all pertinent demographic, historical, clinical, and financial information accurately. As discussed in the previous chapter, because accurate documentation is the legal responsibility of the EMS professional, it must be taken very seriously. It is a solemn duty of any medical professional to document truthfully and accurately. Very few EMS professionals seek to give a false account of patient care. Why, then, is accuracy a problem? Accuracy is a problem when the EMS professional fails to understand the importance of the Patient Care Report as presenting a unified picture of the EMS event. The patient's diagnosis, assessment, treatment, and medical necessity for EMS services must relate and not be in conflict. Accurate PCR documentation presents one clear picture of the EMS professional's management of the EMS event.

Spelling and Grammar

It is vital that spelling and grammar are correct. Other than illegible documentation, nothing impacts the EMS professional's credibility more than spelling and grammatical errors. Many aspects of documentation are not simple, and the subjectivity in documentation will always be present. Errors in spelling and grammar, however, have simple remedies:

- Do not assume you are immune from spelling and grammatical errors. Take precautions to assure structural accuracy in documentation.
- If your electronic PCR has a spell-check function, always use it.
- Purchase a medical terminology pocket guide for assistance in spelling accuracy.
- Proofread every PCR for spelling and grammatical errors.

Correcting Errors or Omissions

Accuracy in PCR documentation is tested when an error or omission occurs. The challenge is not that errors and omissions occur but in the manner in which they are corrected.

Perhaps it has been a long shift. You seemed always to have been in the wrong place at the wrong time and the result is you've had 10 calls in 12 hours. Back at the base, you and your partner are reviewing the PCRs and your partner notes you left out a second dose of a medication. What will you do? You have three options:

1. Ignore it.
2. Correct the omission improperly.
3. Correct the omission properly.

Obviously, if you choose to ignore the omission and go home, you have falsified documentation. Omissions must be corrected properly, and Figure 5-7 provides guidelines for doing so.

Perhaps instead of correcting an error in treatment, you documented something you simply did not intend to write. For instance, you intended to write 5 mg of Valium, but you wrote 10 mg because you were thinking about how it is packaged. Compare the incorrect example to a properly documented error correction in Figure 5-8.

FIGURE 5-7
Guidelines for Correction of PCR Omissions

Because errors in documentation can be interpreted as errors in treatment, the EMS provider must be meticulous when correcting errors. Correcting errors first requires acknowledgment the error has occurred and then a systematic and consistent approach for correction. The following guidelines will assist you in properly correcting errors:

1. Do not attempt to "erase" or completely cross out a documentation error.
2. Instead, draw one line through the error, enter the correction, and add your initials.
3. Add an attachment or addendum to your PCR that notes your name and initials. In doing so, you are saying your initials equal your signature.
4. Enter the date and time of your correction.
5. NEVER create another PCR after a completed PCR has been submitted.

NEVER seek to correct an error or omission if the patient care has come into question. If a lawsuit has been filed, a complaint issued, or a quality investigation initiated, leave it alone. Your attempt to correct a PCR improperly will give the impression that you are trying to change or "cover up" the issue in question. In these situations, the "cover-up" can be worse than the "crime," and something easily explained will be unexplainable if the PCR has been altered improperly.

Incorrect

Medications Administration					
Time	Medication	Dose/Units	Route	Response	Staff ID
1610	NTG	0.4 mg	~~IV~~ SL	▼ Pain	21
Intervention Narrative Comment:					

Correct

Medications Administration					
Time	Medication	Dose/Units	Route	Response	Staff ID
1610	Nitroglycerin	0.4 mg	~~IV~~ SL LDN	BP 112/70, Pain 2:10	21
Intervention Narrative Comment:					
1610: Clarification – 0.4 mg sublingual Nitroglycerin administered. "LDN" = L. Don Nychole					

FIGURE 5-8
Correction of Documentation Errors

EMS Documentation Basic #4—Consistent

The PCR must be consistent because consistency in documentation unifies the demographic, financial, clinical, and transfer of care components of the EMS event. Inconsistencies in documentation will raise questions regarding whether the documentation accurately reflects the patient care. Inconsistency will be the result if the EMS provider views documentation as only recording the EMS event from beginning to end. A common area of inconsistency is the lack of agreement in event times, diagnosis, clinical care, medical necessity, and the narrative. An example of consistency in documentation is shown in Figure 5-9. Note its problems:

- Medication, procedure, and event times are not consistent. Endotracheal intubation is documented as having occurred at 1609, whereas the documented on scene time was 1610. Oxygen is documented as having been initiated at 1615, after endotracheal intubation. Further, administration times for epinephrine and amiodarone mirror the "unit on scene" and "unit left scene" times. When treatment times match with event (dispatch) times, it can suggest that documented times have been "ball-parked" and may be inaccurate.
- The narrative fails to provide both sufficient detail of the assessment and management of the EMS encounter and support for the documented treatment and interventions.
- "En route ACLS continued" was documented, but in reality, no treatment was documented from 1615 to 1635.

The key to consistency is agreement. All aspects of the PCR must agree by presenting a single clear picture of the EMS event.

TIMES:		Incident/Onset Time: UTO
PSAP Call Time: *1600*		Unit Notified Time: *1600*
Unit En Route Time: *1601*		Unit on Scene Time: *1610*
Transfer of Patient Care Time: *1640*		Unit Left Scene Time: *1615*
Patient Destination Time: *1635*		Unit Back in Service Time: *1700*

Medication Administration

Time	Medication	Dose/Units	Route	Response	Staff ID
1610	*Epinephrine*	*1 mg*	*IV*	*None*	*21*
1615	*Amiodarone*	*300 mg*	*IV*	*None*	*21*

Interventions/Procedures

Procedures Performed Prior to EMS Care: *None*

Procedure Authorization: **On-scene** Written Order On-Line *Protocol*

Procedure	Time	Equipment Size	Attempts	Successful	Response	Staff ID
Endotracheal Intubation	*1609*	*7.0*	*3*	<u>*Y*</u> N	*None*	*21*
O2	*1615*	*BVM*	*N/A*	*Y* N		*21*

Narrative
Diagnosis: Cardiac arrest. Arrived on scene, Emergency Medical Responders state patient down one minute post arrival. Amio then Epi given en route. ACLS continued. Last dose on arrival at MCMC–1640.

FIGURE 5-9
Consistency in Documentation

EMS Documentation Basic #5—Concise

concise
Brief and complete.

Documentation must be **concise**. Concise documentation is brief, recording only the essential facts relevant to the diagnosis, medical necessity, clinical care, and treatment. We will discuss principles of concise documentation in Chapter 7.

Summary: Return to Case Study

It is your turn to perform peer review for the new Documentation Compliance program. You're off the streets for two days with over one hundred PCRs to review. You are provided a form to evaluate each PCR using simple criteria regarding whether the documentation is legible and complete. Initially you were dreading the day when your turn would come but you are actually surprised at what you are discovering about documentation quality within your organization. First, documentation among your peers is very inconsistent. Second, there seems to be confusion as to how to be effective in documentation. Third, what really are the company's expectations in documentation? Fourth, while the QI exercise was beneficial, you see the need for the committee to establish standards to evaluate documentation quality. "What is quality in documentation?" you wonder. Unfortunately, your own PCRs are in the stack, and you are troubled that your documentation is often as deficient as that of your peers. You recall the twinge of guilt you often experience when you turn in your PCRs because you know the documentation doesn't quite capture everything about the call. But how can you improve it?

The EMS professional in the case study represents many in EMS today. We know documentation can be better but often do not know where to begin. By providing documentation that is clear, complete, correct, consistent, and concise, you will be well on your way to mastering the skill of EMS documentation.

CHAPTER REVIEW

Review Questions

Please refer to Answers to Chapter Review Questions at the back of this book.

1. List the fundamentals of effective PCR documentation.

2. List three vital functions of PCR documentation.

3. List and describe the characteristics of clear documentation.

4. List the advantages, disadvantages, and principles associated with the use of medical abbreviations.

5. List and describe the characteristics of complete documentation.

6. List and describe the characteristics of correct documentation.

7. List the principles for correcting errors and omissions in PCR documentation.

8. List and describe the characteristics of consistent documentation.

9. List and describe the characteristics of concise documentation.

Critical Thinking

Please refer to Answers to Critical Thinking Discussion Exercises at the back of this book.
1. Discuss how the Five C's of Clinical Documentation impact:

- Professionalism in documentation

- Financing of EMS

- Legal accountability in documentation

Action Plan

1. Apply the Five C's of Clinical Documentation to your daily EMS practice.

2. Assume the role of being your own quality improvement manager. Review your PCRs prior to the end of your shift, applying the Five C's of Clinical Documentation.

Practice Exercises

1. Using the Five C's of Clinical Documentation, evaluate the PCR in Figure 5-10.

   Clear:

   Complete:

   Correct:

   Consistent:

   Concise:

   • What are your general impressions of the PCR?

   • How could this PCR be more effective? Rewrite the PCR using the Five C's of Clinical Documentation.

2. Using the Five C's of Clinical Documentation, evaluate the PCR in Figure 5-11.

   Clear:

   Complete:

   Correct:

   Consistent:

   Concise:

EMS Documentation PCR Example

Incident Location: *Highway 169 – 350th St North*	City: *Rural*
Report Number: *20060004555*	Miles: Start: *00* End: *10*
Incident Date: *5/15/2006*	Medical Control:
Sending Facility: *Emergency Call*	Receiving Facility:*OOH*

Times:

Call Received: *1355*	Dispatch: *1356*
En Route: *1359*	At Scene: *1420*
Depart Scene: *1435*	At Destination: *1452*

Last Name *Tripp*	First Name *John*	Middle

Mailing/Home Address: *6743 South Mulberry Street*

City: *Your Town*	County:	State: *OK*	Zip:

SSN: *000-00-0001*

Age: *47*	Date of Birth: *2/4/1959*

Phone:

Billing Information:
Name/Company:
Insurance Company Address:

Group/ID Number:	Medicare #:	Medicaid #:

Self-Pay: *Yes*

Chief Complaint: *MVC*
Past Medical History: *Irritable Bowel Syndrome*

Current Medications: *Unknown*

Allergies:

Patient Found: *Driver side of pickup truck*

Stretcher Necessary? *MVC*	Reason: *Trauma*

TIME	PULSE	RESP	BP	LOC	SAT		EKG
1420	100	16	120/80	X3	95%		
1430	100	16	120/80	X3	95%	NSR	

TREATMENT:	RESPONSE:

1. *Spinal Immobilization with KED/LSB*
2. *O2 IV and Monitor*
3. *Sager Splint applied to right leg*

Narrative:
Patient found seated in pickup, ran off roadway and struck embankment at approx. 55 mph. Significant damage to front of vehicle. Patient c/o pain to right femur area. Pulses, feeling intact. Past history of irritable bowel syndrome. Has frequent episodes of abdominal pain and diarrhea. States he lost control of vehicle while reaching for a bottle of antacid liquid. Splinted right leg with Sager. Transport to OOH. IV, O2, and Monitor en route.

Crew 1: *A. Medic*, Paramedic	Crew 2: *B. Medic*, Paramedic

FIGURE 5-10
PCR Example

- What are your general impressions of the PCR?

- How could this PCR be more effective? Rewrite the PCR using the Five C's of Clinical Documentation.

EMS Documentation PCR Example

Incident Location: *16524 Ross Road*   City: *Bigtown*
Report Number: *20060006754*   Miles: Start: *00* End: *16*
Incident Date: *6/16/2006*   Medical Control: *Protocol*
Sending Facility: *Emergency Call*   Receiving Facility: *University*

Times:
Call Received: *0200*   Dispatch: *0200*
En Route: *0200*   At Scene: *0210*
Depart Scene: *0225*   At Destination: *0247*

Last Name: *Short*   First Name: *James*   Middle *O*

Mailing/Home Address: *1624 Ross Street*

City:	County:	State:	Zip:
Bigtown	*Common*	*MN*	*55555*

SSN: *000-00-0001*

Age: *66*   Date of Birth: *11/12/1939*

Phone: (612)111-1111

Billing Information:
Name/Company: *Medicare Only*
Insurance Company Address:
Group/ID Number: Medicare #: *000-00-0001B*
Medicaid #:   Self-Pay: *Yes*

Chief Complaint: *Respiratory Distress*
Past Medical History: *COPD*

Current Medications: *Atrovent, Proventil, Singular*

Allergies: *Sulfa*

Patient Found: *Seated on edge of bed in tripod position*

Stretcher Necessary? *Yes*   Reason: *Severe Respiratory Distress*

TIME	PULSE	RESP	BP	LOC	SAT	EKG
0212	*100*	*36*	*90/82*	*X3*	*80%*	*Sinus Tachycardia*
0215	*100*	*32*	*98/76*	*X3*	*85%*	*Sinus Tachycardia*
0220	*120*	*20*	*110/88*	*X4*	*92%*	*Sinus Tachycardia*

TREATMENT:	RESPONSE:
1. *Oxygen @ 15L per Non-Rebreather (0212)*	*Increased saturation to 85%*
2. *Cardiac Monitor (0212)*	
3. *Nebulized Breathing Treatment – Proventil (0215)*	*Increased saturation to 92%*
4. *IV Normal Saline @ TKO*	

Narrative: *Patient found seated on sofa in tripod position. States a 30-minute history of progressive dyspnea, unrelieved by position changes and inhalers. Alert and Oriented X4 w/no neuro deficits. Breath Sounds – Diminished w/wheezes in all fields. Improved with O2 and nebulized breathing treatments.*

Crew 1: *A. Sharp*, Paramedic   Crew 2: *1. Strong*, Paramedic

FIGURE 5-11
PCR Example

CHAPTER 6

Essential Documentation Elements

Key Ideas

Upon completion of this chapter, you should know that:

- PCR documentation is centered on how the entire EMS event was managed.

- Data elements are categories of information obtained for each EMS encounter that provide for organized recording of clinical care and EMS performance.

- There are essential documentation elements for dispatch and demographic data.

- There are essential documentation elements for assessment of the EMS event.

- There are essential documentation elements for EMS treatment and interventions.

- There are essential documentation elements for affirmation of medical necessity and transfer of patient care.

FIGURE 6-1
(Courtesy Tulsa Life Flight/Saint Francis Hospital)

It is 0900 Monday morning, time for another 12-hour shift. It's all about priorities as you arrive at the station and head straight for the coffeemaker before heading out to check the rig. You notice a poster on the bulletin board advertising the arrival of your company's new electronic Patient Care Report and the mandatory in-service training at the end of the month. At the ambulance, your partner and the 2100 night shift are discussing the new PCR, "a product of the Documentation Compliance Committee." Word has it the crew that field-tested the new ePCR found it cumbersome at first but believes it to be a tremendous improvement to the current paper PCR.

Two weeks later, you arrive in the training room for the four-hour in-service. The representative from the EMS software company, an EMS professional himself, begins by relating how the electronic Patient Care Report radically changed his own practice. He explains the purposes that today's ePCRs fulfill in data collection, research, finance, and compliance. Then, he meticulously walks the group through the use of the new ePCR. Although this new device is a shock to your EMS culture, you decide to have an open mind. The first question that comes to your open mind is: "Why does it ask for so much information?"

Questions

Please refer to Answers to Case Study Questions at the back of this book.
1. What are the advantages of using an electronic PCR? What are the disadvantages?

2. Why is the electronic PCR essential to data collection?

Introduction

data elements:
Categories of information obtained for each EMS event that provide documentation of clinical care and EMS performance.

ON TARGET Data drives EMS research. Data is a product of PCR documentation. Therefore, the ability of the EMS provider to capture the EMS event effectively in documentation is one of the keys to the future of EMS.

In Chapter 5 we established the foundation for effective documentation, the how's of PCR documentation. In this chapter we turn our attention to identifying, and appropriately documenting, the essential documentation elements of the EMS event.

PCR documentation must meet data collection requirements. Datasets are subsets of data that describe every aspect of the EMS event. **Data elements** are categories of information obtained for each EMS event that provide for organized recording of clinical care and EMS performance. Data collection is accomplished through extracting information from the data elements. Traditional documentation centered on only recording the care given by the EMS professional. Documentation now focuses on the management of the entire EMS encounter: scene management, demographic and financial data management, clinical management, and transfer of care management.

To facilitate learning and encourage practice of the information presented in the text, the EMS Documentation PCR will be used. This PCR captures the documentation elements, drawing special attention to the essential elements, and reflects current data collection requirements. It is not meant to mimic a paper PCR or any specific ePCR currently on the market. For the purposes of this text, it is a "paper" version of what would be captured electronically.

The EMS Documentation PCR breaks down the EMS event into data categories and is based upon the National EMS Information System (NEMSIS) datasets. The acronym *DATA* frames the necessary data elements of the EMS event:

D = Dispatch and Demographic Elements

A = Assessment of the EMS Event/Patient Elements

T = Treatment and Interventions Elements

A = Affirmation—Medical Necessity and Transfer of Care Statements

DATA summarizes the categories of data that must be recorded in the PCR.

KEY TERMS

Note: Page number indicates where the following key term and definition first appear.

data elements (p. 78)

EMS Event Summary Data

The EMS Event Summary records dispatch information, event times, and the patient's demographic and financial data. The EMS Documentation PCR in Figure 6-2 gives an example of how this information might be recorded. The data elements of the EMS Event Summary include:

- **Agency:** Record the name of the EMS service that responded to the EMS event.
- **EMS Agency Number:** Record the state license number for the service that responded to the EMS event.
- **Incident Number:** Record the number assigned by the agency communications center to the EMS event.

EMS Event Summary			
Dispatch Data			
Agency: *MCEMS*	**EMS Agency Number:** *67890*	**Incident #:** *200600123*	**Unit #:** *3*
Type of Service: *ALS*	**Primary Unit Role:** *Transport*	**Patient Care Report #:** *1237812*	
Pt. Record # *1237812*	**Trauma ID** *None*		
Response Mode:	<u>Emergency</u>	Nonemergency	Other
Incident Location: *13754 150th Street North, Hasketa, OK 79999*			
Response Delay: *None*		**Type:** *Not applicable*	
Transport Delay: *17 minutes*		**Type:** *Cardiac Arrest Interventions*	
Turnaround Delay: *None*		**Type:** *Not applicable*	
Odometer: Beginning *100*	On Scene *105*	Destination *125*	Ending *150*
Complaint Reported by Dispatch: *Chest Pain*			
Crew Member ID: 1. *J. Smith 205* License Level *Paramedic* 2. *I. Kerry 277*		License Level *AEMT*	
Additional Crew Member: *None*		License Level	
Additional Crew Member: *None*		License Level	
Crew Member Role: **Primary Caregiver** *205 / 277*		Secondary Caregiver	Driver

FIGURE 6-2

EMS Documentation PCR—EMS Event Data

- **Unit Responding:** Record the vehicle number unique to the ambulance that responded to the EMS event.
- **Type of Service:** Record the type of EMS service that was requested to respond to the EMS event.
- **Primary Unit Role:** Record the primary role of the unit that responded: transport, nontransport, supervisor, or rescue.
- **Patient Care Report Number:** Many paper PCRs have preprinted numbers associated with the actual PCR. Electronic PCRs generate a PCR number for each incident entered.
- **Essential Element: Response Mode:** Document the mode, emergency, nonemergency, or other, in which you were dispatched to the EMS event.
- **Essential Element: Incident Location:** Record the specific location of the EMS event, including the ZIP code.
- **Response Delays:** If a delay occurred in responding, document the reason. Common examples include weather, traffic, mechanical issues, or being diverted to a different incident. If the delay was a "turnaround" delay, document the reason for such, whether decontamination, restocking of supplies, or transfer of patient care.
- **Essential Element: Odometer Reading:** Record the odometer readings for the responding ambulance at the time of dispatch, on scene, at destination, and at the end. This information is useful in data collection for distance analysis for emergency responses. It is also crucial for appropriate billing of loaded mileage for reimbursement purposes.
- **Complaint Reported by Dispatch:** Often the patient's chief complaint and what is ascertained in the dispatch process are very different from what you ascertain once you arrive at the patient's side. Therefore, it is important to document the chief complaint that is given to you at the time the EMS event is dispatched.
- **Crew Members on the EMS Event:** Record the names and levels of licensure of all crew members assigned to the EMS event. The PCR must be completed by the EMS professional representing the highest level of licensure that provided or directed the care.
- **Crew Member Role:** Record each member's role in the EMS event. Did you provide primary care (you performed the care) or secondary care (you supported another EMS provider or another EMS agency) in the EMS event?

Event Times

The accurate recording of event times is an important task in health care documentation and has tremendous legal implications. Event times include:

- *Public Safety Access Point (PSAP) Call Time*
- *Unit Notified Time*
- *Incident/Onset Time*
- *Unit En Route Time*
- *Unit on Scene Time*
- *Unit Left Scene Time*
- *Patient Destination Time*
- *Transfer of Patient Care Time*
- *Unit Back in Service Time*

Record event times using military time. If you are given event times by your communications center that do not match up with the times you have recorded, work out the discrepancy before you record them in the PCR. If necessary, seek the assistance of a supervisor. The EMS Documentation PCR in Figure 6-3 gives an example of how this information might be recorded.

Event Times	
	Incident/Onset Time: *1530*
PSAP Call Time: *1600*	Unit Notified Time: *1601*
Unit En Route Time: *1602*	Unit on Scene Time: *1610*
Unit Left Scene Time: *1627*	Patient Destination Time: *1635*
Transfer of Patient Care Time: *1640*	Unit Back in Service Time: *1655*

FIGURE 6-3
EMS Documentation PCR—Event Times

Demographic Data

The following demographic information is recorded:

- **Essential Element: Patient's Name:** Record the patient's legal name: last, first, and middle or middle initial.
- **Essential Element: Mailing/Home Address:** Record the patient's legal mailing address.
- **Essential Element: Social Security Number:** Record the patient's Social Security number.
- **Gender, Race, and Ethnicity:** Record the patient's gender, race, and ethnic background.
- **Phone Number:** Record the patient's phone number, including area code.
- **Essential Element: Age, Date of Birth:** Record the patient's age and date of birth. For an infant or newborn, record the age in units, for example, 6 months.
- **Driver's License Number:** Record the patient's driver's license number and the state issuing the license.

The EMS Documentation PCR in Figure 6-4 gives an example of how this information might be recorded.

Financial Information

The following financial information is recorded for payment or reimbursement purposes.

- **Essential Element: Primary Method of Payment:** Record the method of payment for the EMS services. This will be either Medicare, Medicaid, private insurance, personal

Demographic Data			
Last Name *Smith*	First Name *John*		Middle *Q.*
Mailing/Home Address: *13754* *150ᵗʰ Street* *North*	(Street Address)		
City: *Hasketa*	County: *Tulsa*	State: *OK*	Zip: *79999*
SSN: *100-00-0000*	Gender: *Male*	Race: *Caucasian*	Ethnicity:
Age: *65* Age Units: *Not Applicable*		Date of Birth: *12/06/41*	
Phone: *(918)400-0000*		DL # (State): *S56799090 OK*	

FIGURE 6-4
EMS Documentation PCR—Demographics

pay, worker's compensation (WC), or MVC/accident related. The primary method of payment is the payment source that will be billed first or will assume the highest percentage of responsibility for the patient's EMS bill. Keep in mind the patient may have multiple types of payment. For instance, the patient may be covered under Medicare and Medicaid, or have Medicare coverage plus private secondary insurance coverage.

The terms *private pay* and *self-pay* can be confusing but have vastly different meanings to the staff in the billing department. Private pay infers that payment is from a private health plan or insurance company, whereas self-pay infers that the patient will be responsible for payment. Be careful of the manner in which you document private pay and self-pay.

- **Essential Element: Insurance Company Information:** Record the patient's insurance information. This must include the patient or responsible party's insurance ID number, name of the company, mailing address, group ID number, and policy number.
- **Essential Element: Insured Information:** Often, the patient is different from the person who owns the insurance coverage. Common examples are children covered under a parent's health coverage and a patient covered under a spouse's insurance policy. Therefore, it is important to record the name of the person who is insured by the policy and that person's relationship to the patient.
- **Worker's Compensation Information:** If the patient's illness or injury was the result of an incident related to employment, record this information.
- **Name of Closest Relative/Guardian Information:** Record the name and information of the patient's closest relative.
- **Employment Information:** Record the patient's employment information.

The EMS Documentation PCR in Figure 6-5 gives an example of how this information might be recorded.

Assessment Summary Data

The Assessment Summary records the data elements relevant to the EMS professional's assessment of the entire EMS event:

- Assessment of the scene of the incident
- Assessment of the patient's situation, or chief complaint
- Assessment of traumatic injury, if applicable
- Cardiac arrest data, if applicable
- Past medical history
- Physical examination

EMS professionals have the unique responsibility of recording key information that goes beyond the actual patient to the entire event.

Scene Data

- **Public Safety Agencies at Scene:** Record other agencies at scene.
- **EMS Services at Scene:** Record other EMS services at scene.
- **Time of Arrival of Initial Responders:** Record the time the initial responders arrived on scene.
- **Number of Patients:** Record the total numbers of patients involved in the incident. If multiple patients are transported from an EMS encounter, it is important to identify

Financial Information
Primary Method of Payment:
<u>Medicare</u>   Medicaid   Private Insurance   Self-Pay   WC MVC/Accident Related: *Not Applicable*
Secondary Method of Payment: *Blue Cross Blue Shield of OK*
Insurance Company Billing Priority: Primary   <u>Secondary</u>
Insurance Company Address *12345 S. Main Street* City: *OKC*   State: *OK*   Zip Code: *45612*
Insurance Group/ID Name *9077777*   .   Policy ID *100-00-0000*
Insured Information: Last Name: First Name: Middle: *Smith*   *Johanna*   *Q* Relationship: Self   <u>Spouse</u>   Son/Daughter   Other
<u>Work Related:</u>   Y   <u>N</u> Occupational History: *Retired - ISD 111*   Patient Occupation: *Teacher*
Closest Relative/Guardian Information : Last Name: First Name: Middle: *Smith*   *Johanna*   *Q* Street Address: City   State   Zip *13754 150th Street North Hasketa* *OK*   *79999* Phone: (918) *400-0000*   Relationship: *wife*
Patient's Employer: (*Retired*) Address: City: State: Zip:
Patient's Work Telephone: ( )   *Retired*
Reason Information Not Obtained: *Not Applicable*

FIGURE 6-5
EMS Documentation PCR—Financial Information

the patient you are transporting out of the total number from the scene; for example, "1:4" or "2:3." If one or more patients from a multiple patient scene are transported by another ambulance from within your EMS system, note this in the narrative if a specific field is not provided in the PCR. This also applies to patients transported by another agency (such as air medical).

- **Mass Casualty Incident:** Identify whether the incident was a mass casualty incident.
- **Incident Location Type:** Record the type of location.
- **Patient Disposition:** Record the disposition of the patient.
- **Prior Aid:** Record whether the other responding agencies provided aid, including the name of the provider, its title, and the outcome of the care.

If the PCR used by your organization does not provide these specific fields, summarize this information in the narrative section.

Situation Data

Situation data capture the data elements of the patient's chief complaint or presenting problem. When documenting the chief complaint, record what the patient tells you using the

The patient's chief complaint is not the same as the patient's diagnosis.

person's own words. Remember, it's the patient's chief complaint, not yours. Data elements pertaining to chief complaint include:

- **Essential Element: Chief Complaint:** Record the chief complaint, using the patient's own words. Example: "My chest hurts. I feel like a 20-pound weight is on my chest."
- **Duration:** Record the length of time the patient has experienced his or her chief complaint. Example: "The pain began when I was lifting my golf clubs into the car."
- **Time Units:** Record the duration in specific units of time, such as minutes or hours. Example: "30 minutes."
- **Anatomic Location:** Record the physical location for the chief complaint, such as "chest, substernal."
- **Organ System, if Applicable:** If the chief complaint involves an obvious organ system, record this information. Example: "possible cardiac origin."
- **Secondary Complaint:** If present, record any additional complaints ascertained in assessment. Example: "respiratory distress and nausea."
- **Duration of Secondary Complaint:** Record the duration of the secondary complaint. Example: "25 minutes. Began right after the chest pain."
- **Primary Symptom:** Record the patient's primary symptom. Example: "I can't breathe."
- **Other Associated Symptoms:** Record any other associated symptoms that are verbalized by the patient.
- **Essential Element: EMS Diagnosis:** Record your diagnosis of the patient's illness or injury. This is referred to in this text as the EMS Diagnosis, which will be discussed in depth in the next chapter.

The EMS Documentation PCR in Figure 6-6 gives an example of how this information might be recorded. If the PCR used by your organization does not provide data entry fields for this information, summarize it in the narrative section.

Injury Data

If the EMS event is trauma related, the following data elements apply:

- **Intent and Cause of Injury:** Record the intent of the injury. Example: self-inflicted or accidental.
- **Mechanism of Injury:** Record the mechanism that caused the injury. Example: burn, penetrating, or blunt.
- **Vehicular Injury Indicators:** If the patient's injuries are the result of a vehicle incident, it is important for the associated indicators to be recorded, which are critical for research. Example: DOA, rollover, and so on.
- **Area of Vehicle Impact:** If the patient's injuries are the result of a vehicle incident, record the area of the vehicle that received the impact.
- **Use of Occupant Safety Equipment:** Record the use of occupant safety equipment.
- **Air Bag Deployment:** If air bags were deployed in the course of a vehicle incident, record whether front or side bags were deployed.
- **Fall Height:** If the patient's injuries are the result of a fall, record the probable height of the fall.

The EMS Documentation PCR in Figure 6-7 gives an example of how this information might be recorded. If the PCR used by your organization does not provide data entry fields for this information, summarize it in the narrative section.

Scene Assessment Summary	
Scene Data:	
Other Agencies at Scene: *HFD*	Other Services at Scene: *None*
Time Initial Responder on Scene: 1605	Number of Patients at Scene: 1
Mass Casualty Incident Y __N__	Incident Location Type: *residence*

Incident Address:	City	County	State	Zip
13754 150*th* *Street* *North* *Hasketa*		*Tulsa*	*OK*	*79999*

Situation Data:	Prior Aid: *Vital Signs*
Prior Aid By: *Hasketa Fire Department*	Outcome: *Not Applicable*
Possible Injury: *Not Applicable*	

Chief Complaint: *Chest Pain – Dull, "I feel like a 20 pound weight is on my chest."*
Duration of Chief Complaint: *Began while lifting golf clubs*
Time Units: *30 minutes*
Anatomic Location: *Chest, Substernal radiating to left shoulder*
Organ System: *Possible Cardiac Origin*

Secondary Complaint Narrative:
Respiratory Distress, with nausea, began within minutes of onset of chest pain.

Duration of Secondary Complaint:
Time Units: *25 minutes*

Primary Symptom: *Respiratory Distress*

Other Associated Symptoms: *Nausea*

EMS Diagnosis: *Cardiac Arrest*

Secondary EMS Diagnosis: *Chest Pain – Acute Myocardial Infarction*

FIGURE 6-6
EMS Documentation PCR—Scene Assessment Summary

Injury Summary		
Intent of Injury:	__Unintentional__	Intentional/Self
Cause of Injury: *Motor Vehicle Collision*		

Mechanism of Injury:
  Burn   Penetrating   __Blunt__   Other

Vehicular Injury Indicators:
DOA   Fire   Ejection   __> 1 Foot Space Intrusion__   Rollover
Deformity: Side   __Steering Wheel__   Dash   WS

Area of Vehicle Impact:
__Center__ Left Rear   Right Front   Right Side   Left Front   Left Side   Right Rear   RO

Seat Row Location in Vehicle: *Front*   Position of Patient in Seat: *Driver*

Use of Occupant Safety Equipment:
Eye Protection   Lap Belt Protective Clothing   __Shoulder Belt__   Child Restraint
Helmet   Personal Flotation Device   Protective Non-Clothing Gear
None

Airbag Deployment: __None__   Side   Front   Other

Fall: *Not Applicable - MVC*   Height:

FIGURE 6-7
EMS Documentation PCR—Injury Summary

Cardiac Arrest Data

The patient in cardiac arrest requires documentation of additional data elements:

- **When Did the Cardiac Arrest Occur?** Did the arrest occur prior to EMS arrival or after EMS arrived on scene?
- **Estimated Time of Arrest Prior to EMS Arrival:** Record the estimated down time.
- **Etiology of the Cardiac Arrest:** Record the presumed cause of the cardiac arrest. Example: trauma, electrocution, or unknown.
- **Resuscitation Attempted:** Record whether resuscitation was attempted prior to EMS arrival.
- **Arrest Witnessed By:** Record whether the cardiac arrest was witnessed and who witnessed the event.
- **First Monitored Rhythm:** Record the cardiac rhythm first noted upon arrival.
- **Return of Spontaneous Circulation:** Document whether circulation was restored at any time.
- **Resuscitation Discontinued:** If resuscitation efforts were discontinued, record the reason and time for the discontinuation of efforts. Example: "1515, resuscitation discontinued after 15 minutes of ACLS interventions per Mark Smith, Medical Control Physician."

The EMS Documentation PCR in Figure 6-8 gives an example of how this information might be recorded. If the PCR used by your organization does not provide data entry fields for this information, summarize it in the narrative section.

Medical History

Obtaining a patient's medical history is foundational to the practice of the EMS professional. The following data elements are related to the EMS patient's medical history:

- **History Provided By:** Record your source for the patient's history.
- **Sending and Destination Medical Facilities' Record Numbers:** If applicable, record the medical record numbers for sending and destination facilities.

Cardiac Arrest Summary			
Cardiac Arrest Y N After EMS Arrival Y Prior to EMS Arrival <u>N</u>			
Estimated Time of Arrest Prior to EMS: *Not Applicable*			
<u>Cardiac Arrest Etiology:</u> Electrocution Respiratory			
<u>Cardiac</u> Drowning Trauma Other			
<u>Resuscitation Attempted:</u> Not Attempted/DNR			
<u>Defibrillation</u> <u>Chest Compressions</u> Not Attempted—Considered Futile			
Not Attempted—Signs of Circulation			
Arrest Witnessed By: *EMS Staff* First Monitored Rhythm: *VFib*			
Return of Spontaneous Circulation? Y N			
Reason CPR Discontinued: *Not Applicable*			
Time Resuscitation Discontinued: *Not Applicable*			
<u>Narrative Comment:</u> *Witnessed cardiopulmonary arrest @ 1615. Immediate defibrillation @ 150 joules with subsequent return to sinus bradycardia @ 1617.*			

FIGURE 6-8
EMS Documentation PCR—Cardiac Arrest Summary

- **Barriers to Assessment and Patient Care:** Record any barriers to obtaining the patient's medical history or to performing an appropriate assessment.
- **Patient's Primary Practitioner:** Record the name of the patient's primary health care provider.
- **Advance Directives:** If your patient has an advance directive, record the type and whether it was present in the medical records. Advance directives will be discussed in Chapter 13.
- **Allergies:** Record medication, environmental, or food allergies.
- **Medical/Surgical History:** Record the patient's past medical and surgical history.
- **Current Medications:** Record the patient's current medications. The name of the medication, dose, unit, and route must be recorded and whether the medication was taken the day of the EMS event.
- **Presence of Emergency Information Form:** Record whether an emergency information form specific to the patient's health care needs was present.
- **Alcohol/Drug Use Indicators:** Record any indications that suggest the potential for current drug or alcohol use.
- **Pregnancy:** If applicable, record whether the patient is pregnant.

The EMS Documentation PCR in Figure 6-9 gives an example of how this information might be recorded. If the PCR used by your organization does not provide data entry fields for this information, summarize it in the narrative section.

Assessment and Physical Examination

EMS professionals must always conduct and document a complete physical examination. There must be no disagreement between what was done for the patient and what was documented. Take the time to be equally thorough and complete in assessing and documenting each assessment. Document your findings as completely and thoroughly as you performed the assessment and examination.

- **Vital Signs Obtained Prior to EMS Care:** Record vital signs obtained prior to EMS arrival. Example: vital signs obtained by Emergency Medical Responders.
- **Essential Element: Vital Signs:** Record vital signs: blood pressure (systolic and diastolic), heart rate, and respiratory rate. You should obtain, at a minimum, two complete sets of vital signs. If the patient is unstable, vital signs should be monitored at least at intervals of every 5 to 15 minutes, depending on the condition of the patient.
- **Oxygen Saturation:** Record the patient's oxygen saturation level.
- **EKG Rhythm:** Record the patient's EKG rhythm.
- **End-Tidal CO_2:** If applicable, record the patient's end-tidal or other CO_2 measurement and the type of device used.
- **Pain Scale:** Record the patient's pain status. Example: "5:10."
- **Temperature:** Record the patient's temperature and the method used to obtain the reading. Example: "96°F tympanic."
- **Glasgow Coma Scale:** Record the patient's scores from coma scale assessment.
- **Level of Responsiveness:** Record the patient's level of responsiveness.
- **Orientation:** Record the patient's orientation to person, place, time, and event.
- **Glucose Testing:** Record the patient's blood glucose level, sample site, and the device used.
- **Stroke Scale:** If applicable, document the appropriate stroke scale and findings.
- **Apgar Score:** If applicable, record Apgar scoring.

Medical History Summary			
History Provided By: *Patient and Spouse*			
Sending Facility Medical Record # *Not Applicable*			
Destination Medical Record # 456190876			
Barriers to Assessment and Patient Care: None			
Developmentally Impaired		Language	
Physically Impaired		Unconscious	
Speech Impaired		Unattended/Unsupervised	
Hearing Impaired			
Patient's Primary Practitioner:			
Last Name *Oldham*	First Name *Victor*		Middle Name *B*
Advance Directives: DNR Form State/EMS DNR Form			
Living Will Family/Guardian Request (No Documentation) None			
Allergies:			
Medication: *Penicillin, Toprol* (severe bradycardia)			
Environmental/Food: *None*			
Medical/Surgical History:			
Appendectomy 1998 *Angina (6-month history)*			
HTN Since 2000			
Pertinent Family History: *Cardiac – Father had MI @ age 55*			
Social History Narrative:			
Recently retired high school teacher/football coach. Lives with wife in well-kept rural community. Occasional alcohol use in social setting. Nonsmoker – 2 pack per day history up to 1990.			
Current Medications:			
Medication	Dose	Unit	Route
NTG	0.4	*mg metered*	*S L prn*
Diltiazem	30	*mg*	*po QID*
Zoloft	100	*mg*	*po QD*
Presence of Emergency Information Form: Yes No Not Available NA			
Alcohol/Drug Use Indicators: None Smell of Alcohol on Breath			
Patient Admits to Drug Use Patient Admits to Alcohol Use Other			
Pregnancy: Y N Not Applicable Not Available Not Known			

FIGURE 6-9
EMS Documentation PCR—Medical History Summary

- **Trauma Scores:** If applicable, record trauma scores.
- **Narrative Comment:** Summarize the assessment findings with a narrative comment. Narratives and narrative comments will be discussed in the next chapter.

The EMS Documentation PCR in Figure 6-10 gives an example of how this information might be recorded.

Trauma Assessment

Essential documentation elements for traumatic injury include recording assessment findings for each anatomical area. Documenting assessment findings for each anatomical area makes a powerful statement about the care provided by the EMS professional The EMS Documentation PCR in Figure 6-11 gives an example of how this information might be recorded.

Assessment and Physical Exam Summary								
Assessment/Vital Signs	TIME	BP	PULSE	RESP	02 SAT	EKG	PAIN	TEMP
VS Obtained Prior to EMS: *BP:116/80, HR–88, RR–16 @ 1607* Care: Obtained By: *Hasketa FD*	*1611*	*112/60* Method: *Direct*	*90* Rhythm: *Irregular*	*16* Effort: *Labored*	*88%* *See Note*	*NSR w/ PVCs*	*8:10* *10 scale*	*98*
Glasgow Coma Scale Eyes – *4* Verbal – *5* Motor – *6* TOTAL: *15*	*1613*	*80/48* Method: *Direct*	*140* Rhythm: *Irregular*	*24* Effort: *Labored*	*86%*	*ST w/ PVCs*	*8:10*	
Level of Responsiveness: Verbal Unresponsive Alert ◄——— Painful	*1615*	*-0-* Method: *Direct*	*-0-* Rhythm:	*4* Effort:	*80%*	*VF*	-	-
Oriented: Person ◄——— Place ◄——— Time ◄——— Event ◄———	*1617*	*90/60* Method: *Direct*	*60* Rhythm: *Regular*	*20* Effort: *Vent*	*98%*	*Sinus Brady*	-	-
Glucose: Level: *120* Time: *1615*	*1620*	*110/70* Method: *Direct*	*82* Rhythm: *Regular*	*20* Effort: *Vent*	*98%*	*NSR*	-	*97*

FIGURE 6-10
EMS Documentation PCR—Assessment and Physical Exam Summary

Focused Physical Examination

The data elements of the EMS physical examination include:

- **Estimated Body Weight:** Record the patient's actual or estimated weight.
- **Time of Assessment:** Record the time the assessment was completed.
- **General Appearance:** A reference should be made to the patient's general appearance. This will be discussed in the next chapter.
- **Head/Face Assessment:** Record assessment findings from the head and face. Document presence of trauma, deformities, hearing loss, ear or nose drainage, and any asymmetrical findings.
- **Neck Examination:** Record assessment findings from the examination of the neck. Document presence of pain, tracheal position, or presence of jugular vein distention.
- **Chest/Lung Assessment:** Record assessment findings from the examination of the chest, lungs, and heart. Record the general appearance (shape) of the chest and results of chest palpation and auscultation (lungs clear, diminished, crackles, wheezes, rhonchi, or stridor). Describe chest/respiratory movement and expansion, respiratory pattern, use of accessory muscles, and any traumatic findings. If trained, document results of heart tones.
- **Abdominal Assessment:** Record assessment findings from examination of the abdomen. Record the appearance of the abdomen (including presence of scars), shape, tenderness, symmetry, presence of pulsations, referred pain, and auscultation results from four quadrants.
- **GU Assessment:** Record assessment findings from the examination of the genitourinary system. Record presence of pain associated with urination, frequency changes, presence of hematuria, or changes in the characteristics of urine.

Trauma Assessment Summary		
Injury Matrix Skin:		Soft Tissue Swelling/Bruising
Bleeding Controlled	Burn	Dislocation Fracture
Puncture/Stab	Amputation	Bleeding Uncontrolled
Crush	Gunshot	Pain w/o Swelling/Bruising
Injury Matrix Head:		Soft Tissue Swelling/Bruising
Bleeding Controlled	Burn	Dislocation Fracture
Puncture/Stab	Amputation	Bleeding Uncontrolled
Crush	Gunshot	Pain w/o Swelling/Bruising
Injury Matrix Face:	*No findings*	Soft Tissue Swelling/Bruising
Bleeding Controlled	Burn	Dislocation Fracture
Puncture/Stab	Amputation	Bleeding Uncontrolled
Crush	Gunshot	Pain w/o Swelling/Bruising
Injury Matrix Neck:		Soft Tissue Swelling/Bruising
Bleeding Controlled	Burn	Dislocation Fracture
Puncture/Stab	Amputation	Bleeding Uncontrolled
Crush	Gunshot	Pain w/o Swelling/Bruising
Injury Matrix Thorax:	*No findings*	Soft Tissue Swelling/Bruising
Bleeding Controlled	Burn	Dislocation Fracture
Puncture/Stab	Amputation	Bleeding Uncontrolled
Crush	Gunshot	Pain w/o Swelling/Bruising
Injury Matrix Abdomen:	*No findings*	Soft Tissue Swelling/Bruising
Bleeding Controlled	Burn	Dislocation Fracture
Puncture/Stab	Amputation	Bleeding Uncontrolled
Crush	Gunshot	Pain w/o Swelling/Bruising
Injury Matrix: Spine:		Soft Tissue Swelling/Bruising
Bleeding Controlled	Burn	Dislocation Fracture
Puncture/Stab	Amputation	Bleeding Uncontrolled
Crush	Gunshot	Pain w/o Swelling/Bruising
Injury Matrix Upper Extremities:	*No findings*	Soft Tissue Swelling/Bruising
Bleeding Controlled	Burn	Dislocation Fracture
Puncture/Stab	Amputation	Bleeding Uncontrolled
Crush	Gunshot	Pain w/o Swelling/Bruising
Injury Matrix: Pelvis:	*No findings*	Soft Tissue Swelling/Bruising
Bleeding Controlled	Burn	Dislocation Fracture
Puncture/Stab	Amputation	Bleeding Uncontrolled
Crush	Gunshot	Pain w/o Swelling/Bruising
Injury Matrix Lower Extremities:	*No findings*	Soft Tissue Swelling/Bruising
Bleeding Controlled	Burn	Dislocation Fracture
Puncture/Stab	Amputation	Bleeding Uncontrolled
Crush	Gunshot	Pain w/o Swelling/Bruising

Assessment Narrative Comment: *Patient reports falling approximately 5 feet while changing a light bulb. Cause of fall–chair brake/collapsed. Verbalizes pain to cervical area and posterior head, with approximate 3 inch moderate thickness laceration to midoccipital region. Bleeding promptly controlled with sterile dressing.*

FIGURE 6-11
EMS Documentation PCR—Trauma Assessment Summary

- **Back Assessment:** Record assessment findings from the examination of the back. Essential elements of documentation include posture, exact location of pain, presence of radiating pain, and other associated sensations. If back pain is a result of an acute injury, determine whether the patient has urinated since the onset of the injury. If EMS has been called for a patient with chronic pain, determine the factors that have led the patient to call for EMS assistance and record your findings.
- **Extremity Assessment:** Record assessment findings from the examination of the extremities. Record the presence of pain, joint stiffness, rashes or skin changes in the area affected, swelling, fever, and changes to range of motion.

- **Eye Assessment:** Record assessment findings from the examination of the eyes. Record abnormal pupillary response, extraocular motion, and the presence of any visual changes.
- **Neurological Assessment:** Document the presence of any abnormal neurological findings. Record any changes in mentation, memory, weakness, or loss of consciousness.
- **Narrative Comment:** Summarize the assessment findings with a narrative comment.

The EMS Documentation PCR in Figure 6-12 gives an example of how this information might be recorded. If the PCR used by your organization does not provide data entry fields for this information, summarize it in the narrative section.

Focused Physical Exam Summary

Estimated Body Weight: *220 lbs*   **Time of Assessment:** *1611–1630*

General Appearance
Grooming: **Appropriate**   Not Appropriate   Describe:
Gestures/Body Language: Appropriate   Describe: _____
Positioning: *Leaning forward, clutching chest*

Mental Status Evaluation:
Speech: **Normal**   Slurring   Normal/Coherent   Non-Appropriate   Flight of Ideas
Emotional State: **Appropriate**   Depression   Elation   Anger   Other: _____
Thought: **Appropriate**   Anxious   Delusions   Hallucinations   Other: _____
Cognitive Abilities: **Attention Intact**   **Memory Intact**   Deficits _____

Neurological
Seizure Activity: Onset: *None*   Past History   Y   **N**
Generalized: Tonic-Clonic   Absence   Pseudoseizure
Partial: Simple Partial   Complex Partial
Motor: Weakness –   RUE   RLE   LUE   LLE   Generalized
Gait: **Normal**   Abnormal   Describe: _____
Sensory: Pain Status: **8:10**   Location: **Chest**
                 EMS Field Cranial Nerve Testing
I: Smell **Intact**   VIII: Hearing **Intact**
II: Vision **Intact**   IX/X: Swallow Ability **Intact**
III/IV/VI: Eye and Eye Lid Movement **Intact**   XI: Shoulder Shrug **Intact**
V: Facial Muscle Tone **Intact**   XII: Tongue Movement **Intact**
VII: Raise Eyebrows/Show Teeth **Intact**

HEENT
Head: Deformity   Tenderness   Trauma   **No Findings**
Eyes: **Symmetrical**   Asymmetrical   Drainage:
Pupils Left: 2mm   3mm   **4 mm**   5mm   6mm   7mm   Fixed   Reactive   Non-Reactive
Eyes Right: 2mm   3mm   **4 mm**   5mm   6mm   7mm   Fixed   Reactive   Non-Reactive
Ears: Hearing Intact   **Y**   N   Drainage: Y   **N**   Describe:_____
Nose: **Nasal Flaring**   Deviation   Deformity   Tenderness   Drainage   Describe: _____
Mouth/Throat/Neck: Mucous Membranes Moist: **Y**   N   Teeth Intact: **Y**   N
Trachea: **Midline**   Left Deviation   Right Deviation   JVD: Left   Right   **None**

Respiratory Status:
                 Respiratory Pattern
Normal   Bradypnea   Tachypnea   Hyperventilation   Cheyne-Stokes   Kussmaul
                     Apneustic
                 Respiratory Effort
    Symmetrical   Retractions   **Increased Effort**   Use of Accessory Muscles
                     Cough
None   Non-Productive   Productive   Describe: _____
                 Breath Sounds
                 Normal

Decreased BS:	Left	Right	Rhonchi:	Left	Right
Crackles:	Left	Right	Wheezing:	Left	Right

FIGURE 6-12
EMS Documentation PCR—Focused Physical Exam Summary (Continued on p. 92)

<table>
<tr><td colspan="2">

Cardiovascular Status:

Skin Assessment

Warm Dry <u>**Pale Clammy**</u> Cold Cyanotic Mottled Jaundiced Skin

Nail Beds

<u>**Normal**</u> Cyanotic Clubbing

PMI

Location:_____ Thrills Bruits

Heart Tones

<u>**S1 S2**</u> S3 S4 Friction Rub Murmur

Peripheral Edema

4+ = Very Deep/2–6 minutes 3+ = Noticeably Deep – 1 minute 2+ = Disappears in 10–20 seconds 1+ = Slight Pitting–Disappears Quickly <u>**None**</u>

RLE ___ LLE _____ RUE _____ LUE _____ Abdomen_____ <u>**None**</u>

Pulses

<u>4 = Bounding   3 = Full   2 = Expected   1 = Diminished   0 = Absent</u>

Carotid: 2   Brachial: 2   Radial: 2   Femoral: 2

Popliteal: 2   Dorsalis pedis: 2   Posterior tibial: 2

</td></tr>
</table>

GI/GU Status:

Abdominal Appearance : Flat **Round** Distended Symmetrical Asymmetrical

Bowel Sounds: **Present**   Absent   Diminished   Increased

Abdomen LU Assessment–	Guarding	Tenderness	**Normal**	Distension	Mass
Abdomen LL Assessment–	Guarding	Tenderness	**Normal**	Distension	Mass
Abdomen RU Assessment–	Guarding	Tenderness	**Normal**	Distension	Mass
Abdomen RL Assessment–	Guarding	Tenderness	**Normal**	Distension	Mass

Urinary Status: Dysuria Hematuria Polyuria Hesitancy (-)Stream   Incontinence

Female: Discharge STD: _____   *Not Applicable*

Male: Swelling Torsion Lesions   Discharge STD:__ *Deferred – Cardiac Arrest*

Musculoskeletal:

Cervical: **Normal**   Tender/Para-Spinous   Pain to ROM   Tender Spinous Process

Thoracic: **Normal**   Tender/Para-Spinous   Pain to ROM   Tender Spinous Process

Lumbar: **Normal**   Tender/Para-Spinous   Pain to ROM   Tender Spinous Process

RUE: Tenderness/Pain   Redness/Warmth   Weakness   Deformity   Decreased ROM

RLE: Tenderness/Pain   Redness/Warmth   Weakness   Deformity   Decreased ROM

LUE: Tenderness/Pain   Redness/Warmth   Weakness   Deformity   Decreased ROM

LLE: Tenderness/Pain   Redness/Warmth   Weakness   Deformity   Decreased ROM

OB: *Not Applicable*

Gravida: ____ Para: _____ Gestational Age: _____ EDC: _____

Vaginal Bleeding ROM: Y   N   Unknown   Time: _____ Color: _____

Contractions: Onset Time: _____ Timing: Every _____ Duration: _____

Assessment Note:

On arrival found this well-dressed 65-year-old male patient seated in a recliner, leaning forward with right fist clenched over mid sternal region. He reports a 30-minute history of substernal chest pain (8:10) radiating to left shoulder, unrelieved by rest and self-administered NTG spray. Denies fall or other trauma associated with incident.

FIGURE 6-12

EMS Documentation PCR—Focused Physical Exam Summary (continued)

Documentation of patient assessment and examination requires meticulous attention to detail. The difference between a careful assessment that reveals life-threatening illness or injury and "physical exam—no findings" is life and death. If you assess in great detail, you can document in great detail. An excellent patient assessment and examination is dependent upon the EMS professional developing proficient assessment, examination, and documentation skills.

Treatment and Interventions

Essential documentation elements for EMS interventions are specific to the procedure performed. Performing an invasive procedure or administering a medication significantly increases the risk for liability. Therefore, these elements place greater demands on the EMS

EMS Treatment		
BLS Interventions		
Interventions Performed Prior to EMS Care: *Vital Signs Only*		
Intervention Authorization: On-Scene Written Order On-Line *Protocol*		
Airway Management:		
Airway Status: Clear Obstructed Compromised		
Suction: Y N Device Pre-oxygenated Y N		
Time: Device: Notes: _____		
Oxygen Therapy: Device: *NRB* Time: *1610* Flow Rate: *15L*		
Pre-Oxygen Saturation: Post-Oxygen Saturation: *98%*		
Immobilization:		
Spinal *Not Applicable* Extremity		
Time: Time:		
Device: Device:		
Pre-Neuro Status: Pre-Neuro Status:		
Post-Neuro Status: Post-Neuro Status:		
Destination Status: Destination Status:		
Intervention Note: *No oxygen therapy prior to arrival.*		

FIGURE 6-13
EMS Documentation PCR—BLS Interventions

professional's ability to document the procedure appropriately. The data elements for EMS treatment and interventions are:

- **Treatment and Interventions Performed Prior to EMS Care:** This includes interventions provided by Emergency Medical Responders and other EMS agencies.
- **Intervention Authorization:** Record the medical authority for the treatment and interventions. Example: "Protocol or on-line medical control."
- **Basic Airway Management:** Record the airway management used for the patient, including initial management of the airway, oxygen therapy, and the patient's response to therapy.
- **Basic Immobilization:** Record spinal or extremity immobilization, documenting time applied, device used, pre/post neurological exam findings, and reassessment at the destination facility.

The EMS Documentation PCR in Figure 6-13 gives an example of how this information might be recorded.

Airway Procedures

Essential documentation elements for airway procedures include:

- **Specific Procedure:** Record the specific procedure performed.
- **Time of Procedure:** Record the time the procedure was performed. Recording intervention times accurately is essential. Use tools that are already at your disposal such as event markers in the cardiac monitor, or consider using a voice-activated recorder that attaches to your belt like a pager. It is important to have a plan in order to document accurate intervention times.
- **Equipment Used:** Record the equipment used in the procedure such as oral/nasal airway size, laryngoscope blade size, stylet, and endotracheal tube size.

- **Attempts:** Record the number of attempts.
- **Successful:** Record whether the procedure was successful.
- **Response:** Document how the patient responded to the procedure.
- **Staff ID:** Record the ID of the staff member completing the procedure. Document in the narrative the names and levels of licensure of other EMS or medical personnel who assisted in the procedure.
- **Airway Grade:** Record your assessment of the patient's airway, utilizing your agency's tool for assessing a patient's airway prior to intubation.
- **Intubation Adjuncts:** Document any adjuncts that were used to assist in the invasive procedure. Example: gum bougie stylet.
- **Complications/Rescue Airway:** Document any complications and your choice of rescue airway. If complications were encountered in the procedure, document the complications, how you intervened, and the results.
- **ET Tube Confirmation:** Record how placement of the advanced airway was confirmed.
    - Record visualization of the cords (or use of an appropriate intubation assist device).
    - Presence of bilateral breath sounds via direct auscultation of lung fields.
    - Absent air sounds in the epigastric region.
    - Positive chest rise and fall.
    - Expiratory condensation in the ET tube.
    - Confirmation of placement with more than one adjunctive device.
    - Pre/post procedure pulse oximetry.
    - Record the location/position of the ET tube, inflation of the tube cuff, and how you secured the endotracheal tube with an appropriate device. Document verification of tube placement after each time the patient is moved.
    - Document end-tidal CO_2 device used.
- **Destination ET Tube Confirmation:** Thoroughly document how you confirmed placement of the ET tube at the destination facility and how placement was confirmed each time the patient was moved.
- **Narrative Comment:** All invasive procedures require a narrative comment summarizing the procedure. Narratives and narrative comments will be discussed in depth in the next chapter.

The EMS Documentation PCR in Figure 6-14 gives an example of how this information might be recorded.

Cardiac Monitoring and Interventions

Essential documentation elements for cardiac monitoring and interventions include:

- **EKG Lead Monitored:** Record the leads monitored.
- **Multifunction Pads:** Record preplacement of "Hands Off" or multifunction pads.
- **12 Lead EKG Interpretation:** Record the interpretation of the EKG. If serial EKGs were obtained, document the time, interpretation, and changes noted.
- **Defibrillation/Cardioversion:** Record the essential elements for defibrillation: monophasic or biphasic delivery, defibrillation time(s), EKG rhythm, energy level, and staff ID.
- **Transcutaneous Pacing:** Record the essential elements for transcutaneous pacing: rate, energy level, capture, and time initiated.

Airway Interventions						
Procedure	Time	Equipment Size	Attempts	Successful	Response	Staff ID
Suction/OPA	1616	Rigid Catheter #4 Oral Airway	one	Yes		277
Endotracheal Intubation	1617	3 Miller 8.0 cuffed ET	one	Yes		277
Transport Ventilator	1620	------	----	Yes	98% O2 saturation	205/277

Airway Grade: 2

Intubation Adjuncts: Gum bougie stylet

Complications: None   **Rescue Airway:** Combitube

ET Tube Confirmation:
Listed Below:
Auscultation of Bilateral Breath Sounds	Visualization of Tube–Cords
Esophageal Bulb Aspiration	Digital CO2 Confirmation
Waveform CO2 Confirmation	Negative Auscultation over Epigastrium

Destination Tube Confirmation:
Listed Below:
Auscultation of Bilateral Breath Sounds	Visualization of Tube–Cords
Esophageal Bulb Aspiration	Digital CO2 Confirmation
Waveform CO2 Confirmation	Negative Auscultation over Epigastrium

Intervention Narrative Comment:
After successful defibrillation, patient's airway aggressively managed with suction, placement of oral airway, followed by pre-oxygenation and immediate endotracheal intubation. Intubated with 8.0 cuffed ET tube, utilizing a gum bougie stylet, two-person technique. Confirmed endotracheal tube placement per above, inflated cuff with 8-ml of air and secured ET tube with mechanical device. Re-verified placement/tube secure. Patient ventilated with BVM device for three minutes and placed on transport ventilator – adult tidal volume/rate settings. Ventilator settings verified/re-verified. Moved to stretcher – confirmed ET placement, and again @ arrival into ambulance for emergent transport. During transport, patient became responsive to verbal stimuli and followed commands appropriately. Tolerated endotracheal tube through transfer of care.

FIGURE 6-14
EMS Documentation PCR—Airway Interventions

- **Narrative Comment:** All invasive procedures require a narrative comment. Therefore, summarize the intervention with a descriptive comment. Narratives and narrative comments will be discussed in depth in the next chapter.
- **Code Summaries:** Be sure to attach all appropriate rhythm strips and code summaries to the PCR.

Remember: The higher the level of skill required, the greater risk of error, the greater risk of liability, and therefore the greater need for meticulous documentation. The EMS Documentation PCR in Figure 6-15 gives an example of how this information might be recorded.

IV Therapy

Essential documentation elements for IV therapy include:

- **Aseptic Technique:** Record the use of aseptic (or sterile, if appropriate) technique for all invasive procedures.
- **Time:** Record the time the IV was initiated.

Cardiac Monitoring and Interventions					
EKG Lead Monitored: *11, V3*					
12 EKG Time: *1622* Interpretation: *Sinus Rhythm w/acute anterior MI* Time: Interpretation: Time Interpretation:					
Defibrillation: Monophasic *Biphasic*					
Time	Rhythm	Energy Level	Secondary Rhythm	Defibrillation/ Cardioversion	Staff ID
1615	*Ventricular Fibrillation*	*150 joules*	*Sinus Bradycardia*	*Defibrillation*	*205*
Total Number of Shocks Delivered: *1*					
Transcutaneous Pacing: *Not Applicable* Rate: Energy Level: Capture: Time:					
Intervention Narrative Comment: *Delivered one biphasic defibrillation @ onset of ventricular fibrillation via multifunction pads. Immediate return of spontaneous circulation @ 1616.*					

FIGURE 6-15
EMS Documentation PCR—Cardiac Monitoring and Interventions

- **Site:** Record the site that was used to initiate the IV.
- **Catheter:** Record the catheter used in the procedure.
- **Set:** Record the IV infusion set used in the procedure.
- **Attempts:** Record the number of attempts.
- **Fluid:** Record the type of IV fluid.
- **Rate:** Record the infusion rate and total amount infused. If the IV was preestablished by another EMS provider or by a sending medical facility, record the amount of IV fluid in the bag prior to transport.
- **Staff ID:** Record the ID of the staff member initiating the IV.
- **Complications:** Document any complications of the IV insertion.
- **Site Evaluation at Destination:** Document the status of the IV at the time of the transfer of care at the sending and receiving medical facilities, recording the appearance of the IV site pre/post the procedure, indicators of the IV's patency, how the IV was secured, and validation of infusion rate. If an IV was discontinued, it is important to document whether the catheter was intact, a description of the site after discontinuation, and, as always, the use of aseptic technique.
- **Narrative Comment:** Summarize intervention with a narrative comment.

The EMS Documentation PCR in Figure 6-16 gives an example of how this information might be recorded.

Medication Administration

Proper documentation of medication administration is crucial to patient safety. Meticulous care in assessment and examination prior to and after administration is essential. As always, documentation quality should equal patient care quality. Essential documentation elements for medication administration include:

- **Medications Administered Prior to EMS Arrival:** Record medications given by another agency or medical facility.
- **Medical Authorization:** Record the medical control authorization under which the medication was given.

Interventions/Procedures — IV Therapy						
Procedure Performed Prior to EMS Care: *None*						
Allergy Check: *Yes*						
Site Prep: *Betadine Prep Kit*						
Time	Site	Catheter	Set	Attempts	Fluid	Staff ID
1617	*Left Antecubital*	*18 g*	*15 gtt*	*one*	*Normal Saline*	*277*
1630	*Right Forearm*	*18 g*	*Saline lock*	*one*	*-----*	*205*
Successful: *Y* **Complications:** *None*						
Site Evaluation at Destination: *Secured, cool, no redness or signs of infiltration.*						
Maintenance IV: **Site Assessment:** **Fluid:** **Additives:** **Rate:** **Verified in Medical Records:** Y N						
Intervention Narrative Comment: *IV placed immediately after resuscitation. Site prepped, no redness noted prior to venipuncture, using strict aseptic technique. Normal Saline infused @ 50 ml/hr. No redness or infiltration noted. Site cool to touch. Saline lock placed during transport in right forearm. Site prepped, no redness noted prior to venipuncture, using strict asceptic technique. Flushed with 10 ml of Normal Saline w/o infiltration or resistance and secured.* *IV needles recovered and placed in biohazard container.*						

FIGURE 6-16
EMS Documentation PCR—Interventions/Procedures—IV Therapy

- **Allergy Check:** Allergies must always be checked *and* documented prior to the administration of any medication.
- **Time:** Record the time that each medication was given.
- **Medication:** Record the name of the medication, with no abbreviations.
- **Dose and Units:** Record the dose and unit of dosage.
- **Route:** Record the route for the medication.
- **Response:** Document patient responses (positive or negative) to the medication.
- **Staff ID:** Record the ID for the staff member administering the medication(s).
- **Narrative Comment:** Summarize medication interventions with a narrative comment.

Errors in medication administration can be avoided when EMS professionals are diligent about preventing errors. This same effort must go into documenting medication administration so that the accuracy in administration is also reflected in the PCR. See Table 6-1 for guidance in avoiding errors in documenting medication administration. The EMS Documentation PCR in Figure 6-17 gives an example of how this information might be recorded.

Table 6-1 Common Errors in Medication Administration Documentation

1. Abbreviating medications. Avoid using abbreviations when recording the names of medications. Example: MS (morphine sulfate) versus MSO_4 (magnesium sulfate).
2. Improper use of decimal points. Example: a zero must always precede a decimal point, .4 mg versus 0.4 mg.
3. Abbreviating medication doses. Avoid abbreviating *micrograms*, which can be read as *milligrams*, and units, which can be read as a zero if documentation is not legible.
4. Abbreviating routes of administration. Avoid abbreviating subcutaneous (SC), which can be misinterpreted as SL (sublingual) if documentation is not legible.

Interventions/Medication Administration					
Medications Administered Prior to EMS Care: *None*					
Medication Authorization: On-Scene Written Order On-Line *Protocol*					
Allergies Checked: *Yes*					
Medication Administration					
Time	Medication	Dose/Units	Route	Response	Staff ID
1619	*Amiodarone*	*300 mg*	*IV*	*No complications*	205
1627	*Dopamine*	*7 mcg/kg/min*	*IV*	*Stabilized BP (110/70)*	205
Intervention Narrative Comment: *Terminated VFib w/one biphasic defibrillation, followed by amiodarone infusion. No return of VF during EMS care. Medical control physician Ray D. Grey ordered dopamine @ 7 micrograms/kg/ min infused via IV pump.*					

FIGURE 6-17
EMS Documentation PCR—Medication Administration

Safety Interventions

Safety is an important EMS intervention. Therefore, the manner in which EMT personnel ensure the safety of their patients must always be accurately documented in the PCR. Essential documentation elements for safety interventions include:

- **PPE Used:** Record the personal protective equipment used for the EMS encounter and the manner in which it was disposed.
- **Stretcher Safety:** Record how safety was ensured while the patient was on the stretcher. Example: "Patient secured to stretcher with two straps and shoulder harness. Side rails in raised position. Stretcher moved by both crew members at all times. Stretcher secured in ambulance at all times with intact locking mechanism."
- **Patient Belongings:** Record an inventory of any patient belongings and be descriptive. There is a significant difference in recording "watch" and "Rolex brand watch." Whenever you are transporting a patient's personal belongings, document how you secured them for transport.
- **Intervention Narrative Comment:** Summarize safety interventions with a narrative comment.

The EMS Documentation PCR in Figure 6-18 gives an example of how this information might be recorded.

Affirmation

Following the DATA acronym, we have recorded dispatch/demographic data, assessment data, and treatment data. Finally, the EMS professional must affirm:

- Medical necessity (if present) of the EMS services.
- Transfer of care.
- Patient's consent to EMS services.
- Truthfulness and accuracy of the PCR.

Interventions/Safety		
PPE Used: *Gloves, goggles, eye protection/mask*		
Secured to Ambulance Stretcher: *3 straps/shoulder harness*		
Patient Moved to Ambulance:		
Carry	Assisted/Ambulated	Other
Stretcher w/2 crew members	Stair chair	
Patient Moved from Ambulance:		
Carry	Assisted/Ambulated	Other
Stretcher w/2 Crew Members	Stair chair	
Patient Belongings: *Wallet/Gold Fossil Watch* *(given to wife)*		
Secured for Transport: Y N *Not Applicable*		
Intervention Narrative Comment: *Patient placed on long spine board and cervical collar prior to transport and moved to stretcher @ residence, using 6 person lift, taking great care to maintain ET tube. Site surveyed prior to departure: all EMS gloves, disposable supplies, equipment removed from scene/disposed of properly in red biohazard container at Community Hospital.*		

FIGURE 6-18
EMS Documentation PCR—Safety Interventions

Medical Necessity

Essential documentation elements for medical necessity include:

- *EMS Diagnosis*
- *Condition Code*
- *Past Medical History Contributing to This EMS Incident*
- *Bed-Confinement Status*
- *Medical Necessity Not Established*
- *Service(s) Not Available at Sending Facility*
- *Immediate Transport Needed*
- *Medical Necessity for Transport to a Medical Facility of Greater Distance*
- *Sending and Receiving Physicians*
- *PCS Certification Attached*
- *EMS Certification of Medical Necessity Statement*

Appropriate documentation of medical necessity information will be discussed in depth in Chapter 8.

Transfer of Care

Next, the EMS professional affirms that an appropriate transfer of care has taken place. Transfer of patient care is a legal exchange, transferring responsibility of the patient from EMS to the receiving facility. Transfer of care will be discussed in depth in Chapter 12. Essential documentation elements for transfer of care include:

- *Receiving Facility*
- *Patient/Incident Disposition*
- *Transport Mode from Scene*
- *Condition of Patient at Destination*
- *Destination Type*
- *Radio/Phone Report to Destination Facility*
- *Transfer of Care Validation*

Affirmation of Transfer of Care	
Receiving Facility: *Broken Stick Medical Center*	
Patient/Incident Disposition	
Canceled	Dead at Scene
No Patient Found	No Treatment Required
Patient Refused Care	Treated and Released
Treated/Transported	Treated/Care Transferred
Treated/Transported by Law Enforcement	Treated/Transported Private Vehicle
Transport Mode from Scene: *Emergency*	
Condition of Patient at Destination: *Stable post arrest*	
Reason for Choosing Destination Facility: *Closest Appropriate Facility*	
Destination Type: *Level One ED*	
Radio/Phone Report to Destination Facility:	
Time: *1630*	Person: *B. Smith, RN*
Transfer of Care: I have received an appropriate Transfer of Care:	
Time: *1640*	Title: *RN*
Signature: *Brent Smith, RN*	Medical Records Received: *Not Applicable*

FIGURE 6-19
EMS Documentation PCR—Affirmation of Transfer of Care

The EMS Documentation PCR in Figure 6-19 gives an example of how this information might be recorded.

The Narrative

The narrative provides a summary of key information relating to the EMS event and will be discussed in depth in the next chapter.

Signatures

The PCR is a legal document that must be signed by the patient and the EMS professionals providing the care and transport. EMS systems differ in the information that is placed in the patient signature field. Generally a patient signing the PCR is affirming consent to treatment and transport, release of medical information, and authorization for billing purposes. The following information should be added to patient signature areas.

- Receipt of any patient information provided by EMS, such as a copy of the PCR and medical records
- Receipt of any patient belongings given directly to the patient

  It is not always possible to obtain a patient signature. Examples include:

- The competent patient that does not sign his or her name, but uses an *X* or other mark. You must have a witness who validates the mark being the patient's signature. Obtain the person's name, address, and relationship to the patient, and document the reason for the nontraditional signature.
- The deceased patient. Record "patient deceased" in the signature field.
- The patient who is unable to sign for his or her care. The person who is legally authorized to sign for the patient should write the patient's name on the signature line, and add "BY" followed by his or signature and printed name. Example: "John Smith BY *June Smith*, June Smith (Power of Attorney)." Follow local guidelines when presented with this situation. If available, include a hard copy of the patient's power of attorney document.

Legal Signatures				
Patient Signature: John Smith BY *Johanna Smith (Patient post cardiac arrest/intubated on ventilator)*				
Belongings Received: *None/Given to wife prior to transport*				
Staff Signatures:				
Signature:	*James Smith*	James Smith	License #	*P89999999*
Signature:	*Ivan Kelly*	Ivan Kelly	License #	*1 999999999*
Peer Reviewed:	*Y*	N		

FIGURE 6-20
EMS Documentation PCR—Legal Signatures

The PCR is complete with your signature, attesting to the truthfulness and accuracy of the information you have recorded. The EMS Documentation PCR in Figure 6-20 gives an example of how this information might be recorded.

Summary: Return to Case Study

A month has passed since the implementation of the new ePCR. Originally, you were concerned with the amount of data entry it required. However, you have acclimated over time and now you wouldn't think about going back to documenting on paper for a number of reasons. First, the PCR essentially is built for you as it leads you through the datasets. Second, you don't have to be concerned about having PCRs returned to you any longer "because the billing department can't read them." Third, the PCR has a built-in "quality manager" that won't allow you to exit out of the PCR until all required documentation fields have been completed. You are beginning to feel better about EMS documentation, except for writing the narrative.

Effective documentation must include specific essential information that is obtained throughout the EMS encounter. In this chapter we followed the acronym, **D** (Dispatch and Demographic Elements), **A** (Assessment of the EMS Event/Patient Elements), **T** (Treatment and Interventions Elements), **A** (Affirmations—Medical Necessity and Transfer of Care Statements) as a guide for capturing the essential data elements for the EMS event.

CHAPTER REVIEW

Review Questions

Please refer to Answers to Chapter Review Questions at the back of this book.

1. List the four data collection categories.

2. List the essential documentation elements for reimbursement.

3. Describe the difference between diagnosis and chief complaint. What is the importance of this difference in documentation?

4. Which of the following represents a chief complaint and which represents a diagnosis?

    - Multiple system trauma
    - Fall
    - G tube replacement

    - Hip fracture
    - Headache
    - Knee pain

5. List the critical documentation elements for IV therapy.

6. List the critical documentation elements for medication administration.

7. List the critical documentation elements for airway management.

8. List the critical documentation elements for patient safety.

9. List the critical documentation elements for transfer of care.

Critical Thinking

Please refer to Answers to Critical Thinking Discussion Exercises at the back of this book.

1. What is the difference between documenting "head-to-toe exam—no findings" and documenting positive or negative findings for each anatomical area?

2. Why is it important to document care performed by other health care providers prior to your arrival, such as spinal immobilization, airway management, IV therapy, and medication administration?

Action Plan

1. Investigate the various types of pain assessment tools that might be applicable to your EMS practice.

2. Investigate the various types of airway management assessment tools that might be applicable to your EMS practice.

3. Investigate ways to improve your history-taking skills.

Practice Exercises

1. Using the essential elements for spinal immobilization, document a narrative example.

Narrative Snapshot

2. Using the essential elements for oxygen therapy, document a narrative example.

Narrative Snapshot

3. Using the essential elements for transfer of patient belongings, document a narrative example.

Narrative Snapshot

4. Using the essential elements for IV therapy, document a narrative example.

Narrative Snapshot

5. Using the essential elements for endotracheal intubation, document a narrative example.

Narrative Snapshot

CHAPTER 7

Narrative Documentation

Key Ideas

Upon completion of this chapter, you should know that:

- The EMS professional's ability to document a narrative determines the ultimate quality of PCR documentation.

- Proficiency in narrative documentation determines the effectiveness of the EMS professional's documentation.

- SOAP, CHARTE, Head-to-Toe, and Review of Systems are common narrative formats, but each fails to summarize the entire EMS encounter adequately.

- The ultimate test of documentation quality rests upon the effectiveness of the narrative in summarizing key essential elements of the EMS encounter.

- The Focused EMS Event Summary ties together the essential elements for summarizing the EMS encounter.

- The EMS Diagnosis is the medical conclusion made by the EMS professional as a result of EMS assessment and examination.

- Procedure notes capture essential information specifically relating to interventions.

FIGURE 7-1
(Courtesy Lakes Region EMS, North Branch, MN)

CASE Study _____

You and your partner have finished a particularly difficult transport involving a critically ill patient. This call was uniquely different from most of the routine EMS encounters you respond to. First, a number of things went wrong. Dispatch didn't send the closest unit; you waited while your partner got a coffee, extending your response time; medical control wasn't available; and you were diverted five minutes out from the receiving hospital. Most of all, you were significantly challenged by the complexity of the patient's history, medical condition, and the amount of interventions that were required. Everything about the patient care made you think. The patient's history, medications, and even allergies challenged you to think critically through the diagnosis and treatment process. Now, sitting in the report room you face another challenge. How will you adequately summarize this EMS encounter? So much of this call, you reason, simply does not fit into any of the documentation fields in your EMS system's PCR. Thankfully, you have the narrative to tell the whole story of this encounter.

Questions

Please refer to Answers to Case Study Questions at the back of this book.

1. What is the purpose of the narrative in EMS documentation?

2. What are the dangers of summarizing the entire EMS encounter in the narrative section?

Introduction

The narrative first appeared in PCRs in the 1970s and began serving an important function in EMS documentation. The narrative section allowed pioneering EMS professionals to describe the EMS encounter and how they cared for the patient and provided a voice for professional self-expression. As EMS advanced, the narrative became a vital documentation tool and a fixture in the Patient Care Report. The **narrative** in EMS documentation refers to free text areas in the PCR (paper or electronic) in which the EMS professional can document anything.

Although the narrative is universal to virtually all PCRs, its use and application are not. Some EMS professionals view the narrative as nonessential, whereas others consider it the place to re-create the EMS encounter. The narrative sections of PCRs can look as different as the many EMS professionals writing them.

Because EMS documentation is mostly about the narrative, narrative documentation serves and represents the most important function in EMS documentation. The purposes of the narrative are:

- Ties the EMS event together, allowing for highly descriptive information of the EMS event.
- Affirms the medical necessity of the services provided.
- Educates the reader as to why and how the EMS professional intervened. Remember, not everyone reading a PCR will have an EMS background.
- Explains exceptions—issues or events that require further information.

narrative
Open text field(s) that allow for recording of information deemed important by the clinician.

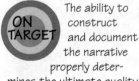

The ability to construct and document the narrative properly determines the ultimate quality and effectiveness of the documentation.

The purpose of the narrative is not to tell the story, provide a complete summarization, or offer a play-by-play account of the EMS event.

- Reinforces safety interventions.
- Provides a place for documentation when specific fields are not provided in the PCR, such as a detailed physical examination.

Other documentation fields simply record information, but the narrative serves to strengthen (or weaken) the appropriateness of the EMS professional's management of the EMS event by summarizing key information. To be truly effective in EMS documentation, you must be proficient in narrative documentation. This chapter will focus on the essential elements of narrative documentation. The Focused EMS Event Summary will be introduced as a guide to attaining proficiency in narrative documentation.

KEY TERMS

Note: Page numbers indicate where the following key terms and definitions first appear.

narrative (p. 106)　　　　　　EMS Diagnosis (p. 112)　　　　　procedure notes (p. 112)

Focused EMS Event Summary
(p. 111)

Narrative Fundamentals

The narrative is foundational to health care documentation. Applying the Five C's of Clinical Documentation to data fields in a paper PCR is somewhat easy, and sometimes the only choice is whether or not to leave the field blank. Many of today's ePCRs make PCR completion almost mindless by requiring mandatory data entry in required fields. Tools such as menus and drop-down boxes make much of PCR completion a cinch. However, it is important to understand that data entry is not EMS documentation and that the PCR is not complete when all required fields have an entry. Instead, it is complete when all appropriate information is recorded and the EMS event has been appropriately summarized. Because this cannot be done independently from the narrative, the narrative really makes or breaks the EMS professional's documentation.

The fundamentals of writing a narrative include the following:

1. The narrative is written to provide descriptive detail of essential elements of the management of the EMS event. The goal of narrative documentation is not to re-create the EMS encounter or provide a play-by-play account but rather to provide descriptive detail. By pulling out the key elements in assessment, examination, and interventions, descriptive documentation paints a clear, concise word picture. Consider the correct example in Figure 7-2 of a diagnosis statement from a narrative. See Table 7-1 for additional examples of descriptive narrative statements.

2. Narrative documentation must be focused in order to fulfill a specific purpose and to target certain areas of the EMS encounter. Consider the two examples in Figure 7-3. Lack of focus in narrative documentation, a common error, allows for different inter-

Narrative Snapshot
EMS Diagnosis: *Parkinson's disease: patient ambulates with difficulty with a shuffling gait, stooped posture, and dragging of right foot.*

FIGURE 7-2
Narrative Example: Diagnosis Statement

Table 7-1 Descriptive Narrative Statements

Nondescriptive	Descriptive
Patient complaining of weakness	This usually ambulatory 67-year-old female patient has been bed-confined for 24 hours due to severe nausea, vomiting, and diarrhea.
CVA	Patient presents with no changes to primary history of CVA (1 year post onset): total right hemiplegia and slurred speech.
Alert and Oriented X4	Patient is alert and oriented to person, place, time, and event, appropriately answering all questions.
Patient found complaining of pain to right knee	Patient found in right recumbent position, with right lower extremity angulated under left leg, complaining of severe (10:10) pain to right knee after fall from top of 6' stepladder.
Patient presents with chest pain	Chest pain (30 minutes since onset), described as "it feels like a truck is parked on my chest," with dull radiating pain to left jaw.

pretations, conclusions, and assumptions about the EMS event. The solution is the Focused EMS Event Summary, which will be introduced later in the discussion.

3. Professionalism is absolutely essential in narrative documentation. Apply the Five C's of Clinical Documentation to narrative documentation:

- A clear narrative is legible and understandable, leaving the reviewer with only one interpretation.
- A complete narrative appropriately summarizes all essential elements of the EMS professional's management of the EMS encounter.
- A correct narrative provides a factual and truthful summarization of the EMS professional's management of the EMS encounter.
- A consistent narrative ties together in agreement the EMS Diagnosis, chief complaint, assessment, treatment, safety interventions, medical necessity, and transfer of care.
- A concise narrative summarizes only the relevant facts of the EMS encounter.

Incorrect

Narrative Snapshot
Patient presents with abdominal pain. History of asthma, depression, and diverticulitis in 1978. Dr. Kinney is the patient's MD. Found the patient seated. Pain mid-abdomen. Denies chest pain or SOB. No meds or allergies. Treatment O2, IV, and monitor. Ate today. Transport to hospital.

Correct

Narrative Snapshot
Patient presents with two-day history of severe right lower quadrant abdominal pain, radiating to umbilical area. Patient reports pain progressive from 2:10 yesterday to 8:10 on EMS arrival. Nausea (12 hours since onset), with 2 episodes of vomiting in last 4 hours.

FIGURE 7-3
Narrative Example: Focusing the EMS Narrative

Narrative Snapshot
This 46 year old belligerent male patient was found in custody of police department, obviously intoxicated.

Correct

Narrative Snapshot
Called to police department to evaluate a 46 y/o male patient. Officer Smith reported they transported patient to their station after altercation (with police officers) while attempting to serve an arrest warrant. Patient verbalizes no chief complaint, pain, or discomfort. However, a 2-inch partial thickness laceration noted to forehead above right eye. Patient uncooperative with attempts to assess and examine. Noted alcohol-type odor when patient spoke.

FIGURE 7-4
Avoiding the Commentary in Narrative Documentation

Narrative Snapshot
John Deux, John Deux, NREMT-I, I-123456

FIGURE 7-5
Narrative Example

The narrative must also be free of subjective information or personal opinion. The EMS professional must resist the temptation to use the narrative for commentary regarding any aspect of the EMS encounter. See Figure 7-4 for an example of an incorrect narrative statement that reflects personal opinion and bias, along with a corrected example.

4. Appropriate medical terminology must be used. By documenting in the universal language of health care, the narrative will reflect professionalism.

5. One function of the narrative is to educate. The physician, emergency nurse, and those within your EMS system will most likely understand the EMS care; however, those working outside EMS, such as for reimbursement purposes, may not. Therefore, the narrative must be written in an organized fashion that leads to an accurate conclusion regarding the EMS services provided.

6. Narrative documentation must be signed, with signature, printed name, and title. Consider the correct example in Figure 7-5.

Traditional Documentation and Narrative Formats

Narrative formats provide focus and structure to documentation and assist the author in staying on course. Numerous narrative formats are available in PCR documentation. Some EMS systems mandate the use of a particular format, whereas others leave the decision to the individual EMS provider. The most commonly used formats are the SOAP (SOAPIER), CHARTE, Head-to-Toe, and Review of Systems formats.

SOAP Format or SOAPIER

The SOAP/SOAPIER is perhaps the most commonly used EMS documentation format:

S = Subjective Data (information reported by the patient or family member, documented in the manner in which the patient/family member stated it = symptoms)

Narrative Snapshot

S—"I hurt my back playing basketball. I went for a rebound and felt a pop right here in my lower back (patient referred to lumbar region @ approximately L5). I have excruciating pain in my lower back that shoots down my left leg. Please don't move me; I don't think I can stand it." States pain knifelike with the feeling of electricity referring down posterior left lower extremity.

O—Found patient prone on cement basketball court, representing position of comfort. Pain scale—10:10. Guarding – does not want to be moved due to aggravation of pain. Denies fall subsequent to injury. Exam and vital signs per flowchart. Remained alert, oriented to person, place, time, and event all times during care.

A—Athletic well-conditioned male with signs indicative of possible lumbar disc herniation. Head-to-toe trauma exam = no findings. Neuro exam = Motor/Sensory intact all extremities with increased pain with movement.

P—Spinal immobilization and pain management.

I—Immediate spinal immobilization. Placed on long spine board. Head immobilized. Pain management with 4 mg morphine sulfate.

E—Patient states, "Pain is much worse on this hard board. You've got to get me off this thing now." Pain decreased to 5:10 within five minutes after morphine.

R—No changes in condition during transport.

FIGURE 7-6
Narrative Example: SOAPIER Narrative Format

O = Objective Data (observations made by the EMS professional, including vital signs or other diagnostic measurements = signs)

A = Assessment Data (interpretations and conclusions made in response to the subjective and objective data = diagnosis)

P = Plan (treatment plan for the patient)

I = Interventions (actual interventions)

E = Evaluation (evaluation of the patient's response to the plan and interventions)

R = Revision (changes in care based upon evaluation of the care)

This format is problem oriented, focusing on the patient's chief complaint or presenting problem. Figure 7-6 provides an example of an EMS narrative using the SOAPIER format.

CHARTE Format

The CHARTE format is similar to the SOAPIER format but includes information relating to the EMS transport:

C = Chief Complaint
H = History of Present Illness
A = Assessment
R = (Rx) Treatment
T = Transport
E = Evaluation

This method has been adopted as the format of choice by many EMS system medical directors. Figure 7-7 provides an example of an EMS narrative using the CHARTE format.

Head-to-Toe Format

The Head-to-Toe Format narrative format is the oldest format used in EMS narrative documentation and it's easy to understand why. Because EMS providers have traditionally been trained to assess in a head-to-toe fashion, it's also reasonable to document in a head-to-toe

<table>
<tr><td colspan="1" align="center">**Narrative Snapshot**</td></tr>
</table>

C–Severe headache, described as sharp pain to frontal area, "I've never had a headache like this before."
H–Sudden onset 30 minutes prior to calling EMS. No prior history of headaches of this severity. Patient has a 2-year history of hypertension, medicated with Lopressor.
A–45-year-old female found supine on sofa at residence. Alert and oriented to person, place, time, and event with no motor or sensory neurological deficits. Denies chest pain, respiratory distress, or other symptoms. Head-to-toe examination—no pertinent findings.
R–Oxygen, monitor. Requested order for morphine for pain. Denied by medical control.
T–Transported in position of comfort, secured to stretcher.
E–No changes.

FIGURE 7-7
Narrative Example: CHARTE Narrative Format

<table>
<tr><td align="center">**Narrative Snapshot**</td></tr>
</table>

62-year-old male patient found supine on ground. Fell approximately 10 feet from ladder while painting his house. Neuro: Alert and oriented with no deficits. Head: Small laceration, approximately 1 inch/partial thickness to posterior head. Neck: No findings. Chest: No bruising, crepitus, or pain on palpation. Breath Sounds: – Equal/clear bilateral breath sounds with no distress. Abdomen: Soft, non-tender. Extremities: Deformity to right mid-shaft femur area. Pulse: Sensory intact. Motor: Able to move toes.

FIGURE 7-8
Narrative Example: Head-to-Toe Narrative Format

fashion. Figure 7-8 provides an example of an EMS narrative using the Head-to-Toe documentation format.

Review of Systems Format

The Review of Systems (ROS) or Body Systems format is used in some form by most physicians. Because the Review of Systems organizes examination data in a general head-to-toe sequence, noting pertinent findings for each organ system, it represents the most targeted and detailed documentation format. Figure 7-9 provides an example of an EMS narrative using the Review of Systems format.

The SOAPIER, CHARTE, Head-to-Toe, and Review of Systems narrative formats each have obvious strengths. However, all these narrative formats fail to summarize the essential elements unique to the needs of the EMS professional.

The Focused EMS Event Summary

Focused EMS Event Summary
A documentation format that summarizes the EMS event by focusing documentation on critical elements of the EMS event.

The ultimate test of documentation quality rests upon the effectiveness of the narrative in summarizing key essential elements of the EMS event. The narrative formats common to EMS, while serving the needs of physicians, fail to summarize the EMS event and are not well suited for use in EMS documentation. The SOAP/SOAPIER and CHARTE formats limit their focus to the patient's chief complaint, assessment, and treatment. The Head-to-Toe and Review of Systems formats focus upon the physical examination. However, the **Focused EMS Event Summary** is designed to capture the critical information that will appropriately summarize the EMS event. The essential elements of the Focused EMS Event follow.

Narrative Snapshot

General Statement: 35-year-old female presents with mild respiratory distress, stating a three-day history of sore throat and fever. VS: per flowchart. Temp 102.5.

Appearance: Well-nourished female found supine in bed. Numerous OTC cold preparations found at bedside.

Skin: Hot/Dry with poor skin turgor. Mild tenting of skin. Mucous membranes dry. Nail beds pale.

Hair: Dry/matted appearance.

Mental Status: Alert and oriented to person and place only. Did not know day and time, otherwise memory appears intact. Attention span is delayed.

Neuro: Weakness—states unable to walk due to dizziness. Motor and sensory functions intact.

HEENT: No trauma or deformities. PEARL, no visual changes. Hearing intact with no drainage. Almost continual green nasal discharge.

Chest/Lungs: Mild dyspnea, chest expansion, and excursion normal.

Breath Sounds: Expiratory wheezes.

Productive cough: Green, blood-tinged sputum.

Cardiovascular: Denies chest pain, discomfort, or palpitations. Rate and rhythm regular. No edema.

GI: Denies abdominal discomfort. Abdomen rounded, symmetrical. Bowel sounds diminished. No pain or tenderness.

GU: Denies flank pain, pain, or hematuria. Last urine output – 10 hours ago.

Endocrine: States a recent 10-pound weight loss.

MS: Denies joint pain, range of motion intact, strength and muscle tone intact.

FIGURE 7-9
Narrative Example: Review of Systems Narrative Format

EMS Diagnosis
The medical conclusion (diagnosis) made by the EMS professional as a result of EMS assessment and examination.

If your EMS system's PCR does not provide data fields allowing for a thorough medical or trauma assessment, the EMS professional should incorporate the Review of Systems format into the Focused EMS Event Summary.

procedure notes
Notes that capture essential information relating to interventions.

1. **EMS Diagnosis:** The Focused EMS Event Summary begins with, and is built upon, the EMS Diagnosis. The **EMS Diagnosis** is the medical conclusion (diagnosis) made by the EMS professional as a result of EMS assessment and examination. The EMS Diagnosis is not, and should not be confused with, the medical diagnosis made by the physician. The EMS Diagnosis will later assist in determining the appropriate condition code for the billing process.

The outdated idea that EMS professionals "do not diagnose" must be discarded. Although this was an appropriate reflection of EMS in its early days, this long-embraced paradigm now undermines the EMS profession. As an EMS professional you diagnose to your level of licensure and provide interventions based upon your diagnosis. The EMS Diagnosis recognizes this fact of EMS practice and represents the foundation for the Focused EMS Event Summary.

2. **Medical Necessity Statement:** The Medical Necessity Statement, discussed in detail in the next chapter, provides the EMS professional a uniform structure for documenting medical necessity.

3. **Scene Summary:** Summarize key findings from the scene relating to the patient's diagnosis, such as mechanism of injury and the patient's presentation.

4. **Review of Systems:** Summarize key findings from the physical examination relating to the EMS Diagnosis for which you provided an intervention. If your EMS system's PCR does not provide data fields allowing for a thorough medical or trauma assessment, the EMS professional should incorporate the Review of Systems format into the Focused EMS Event Summary. Figure 7-10 lists the components of the EMS Review of Systems.

5. **Intervention Summary:** Summarize interventions and the patient's response to the interventions by documenting a procedure note. **Procedure notes** capture essential information relating to interventions, for example, the initiation of an IV, spinal immobilization, or a safety intervention such as placing a patient in restraints. Every invasive procedure must have a procedure note that summarizes the intervention. Ideally, PCRs should provide space for procedure notes immediately following the procedure or intervention. If your EMS system's PCR does not enable you to document in this manner, you

```
General/Appearance
Social History
Past Medical History
Mental Status
Neuro
Integumentary
HEENT
Respiratory
Cardiovascular
Gastrointestinal
Genitourinary
Musculoskeletal
```

FIGURE 7-10
EMS Review of Systems

Narrative Snapshot

Procedure Note: IV access obtained, using aseptic technique in left forearm with 18 gauge Angiocath. Saline lock attached and flushed with 5 ml sterile IV solution without resistance or infiltration. Site cool, with no redness or pain before and after procedure. Site secured with IV membrane.

FIGURE 7-11
Narrative Example: Procedure Note

Narrative Snapshot

Hospital bed placed in lowest position. After informing patient, lifted from hospital bed to stretcher using 4 staff members. Stretcher rails raised and patient secured to stretcher with 3 straps and shoulder harness without incident. Evaluated safety integrity prior to moving stretcher.

FIGURE 7-12
Narrative Example: Safety Summary

---

must include this information in the narrative. Consider the correct example of a procedure note in Figure 7-11.

6. **Changes/Response to Changes:** Summarize any changes in the patient's condition and how you intervened in response to changes. If your patient remained unchanged during the course of your care, this must be referenced in the summary.

7. **Safety Summary:** Summarize the safety interventions provided for the patient at three critical stages of the EMS event: on EMS arrival, during transport, and at destination/transfer of care. See the example in Figure 7-12.

8. **Disposition Summary:** Summarize key information relating to the transfer of care, which will be discussed in Chapter 12.

Figure 7-13 provides an example of the Focused EMS Event Summary. The Focused EMS Event Summary summarizes, around the EMS Diagnosis, all aspects of the EMS event, beginning at the scene and ending with the transfer of care.

Focused EMS Event Summary

<u>EMS Diagnosis</u>: *Acute Myocardial Infarction/Substernal Chest pain radiating to left shoulder, unrelieved by rest.*

<u>Medical Necessity Statement</u>: *Acute Myocardial Infarction, confirmed by12-lead EKG, requiring emergent transport to closest hospital with emergency interventional cardiac services for possible lifesaving interventions.*

<u>Scene Summary</u>: *Emergency Medical Responders present, managing patient appropriately on arrival. Assisted with movement and transport of patient to ambulance, retrieving all equipment and supplies from residence prior to departure.*

<u>EMS Review of Systems</u>:
<u>General/Appearance</u>: Adult male leaning forward while seated on edge of sofa in elegant, well-kept surroundings. Clutching chest, stating, "It feels like a truck is parked on my chest." Pain Scale – 9:10.
<u>Social History</u>: Wife at side, stating "He's been under a lot of stress since the bankruptcy last month, his blood pressure is off the charts, and he's now smoking three packs a day and going through a pint of bourbon every two days."
<u>Past Medical History</u>: Denies diagnosis of hypertension or medication use.
<u>Mental Status</u>: Appropriate dress, behavior–agitated. Oriented to person, place, time, and event. Noted difficulty in speech articulation, responding to questions only with short/appropriate answers due to physical discomfort.
<u>Neuro</u>: Testing of field applicable cranial nerves–intact.
<u>Sensory function</u>: Touch/pain intact bilaterally.
<u>Motor</u>: Minimally tested–moves all extremities/not encouraged.
<u>Integumentary</u>: Skin is pale, cool, and diaphoretic. Nail beds pale.
<u>HEENT</u>: Head no trauma, vision/hearing intact. Trachea midline with no tugging.
<u>Respiratory</u>: Denies dyspnea. Chest movement and expansion symmetric. Respirations rapid. Breath sounds clear all fields.
<u>Cardiovascular</u>: Desires to keep clutched fist over left chest. Neck veins flat. Heart tones S1 and S2 audible. No peripheral edema. Strong, non-bounding pulses present and equal all extremities.
<u>Gastrointestinal</u>: Abdomen soft, rounded, free of pulsations. Bowel sounds present all four quadrants. Denies pain, referred pain, or rebound tenderness.
<u>Genitourinary</u>: No changes or findings.
<u>Musculoskeletal</u>: No traumatic findings. Muscles and extremities symmetric without obvious deformity. Movement not encouraged.

<u>Intervention Summary</u>: *Immediately intervened per standing Metro East Clinical Guidelines: Applied high-flow oxygen, IV placed, secured, and monitored, sublingual nitroglycerin administered, EKG monitored, 12-lead EKG acquired and confirmed anterior MI with 2 mm ST elevation in V3-V4. Sent via cellular to medical control. STEMI protocol activated, with arrival at ED in under 30 minutes. Additional medications administered en route with no side effects. Reference procedure notes and flowcharts for specific intervention details.*

<u>Change in Patient Status</u>: *None*

<u>Safety Summary</u>: *Not allowed to move self or ambulate at any time, gently moved to stretcher at residence with 4 person lift using proper body mechanics. No increased discomfort or change post movement. All items secured for transport. Moved to gurney at ED with slider board, 4 persons assisting this crew.*

<u>Disposition Summary</u>: *Explained transfer of care and STEMI procedures to patient prior to arrival. No delays in transfer of care: 1515 to R. Bean, RN. Brown wallet and gold Rolex branded watch given by patient to wife at time of care transfer. Keith O'Keefe,* Keith O'Keefe, Paramedic.

FIGURE 7-13
Focused EMS Event Summary Example

Summary: Return to Case Study

Completing the PCR, you make two copies, one for the physician and one for the RN to place into the patient medical record. You quietly hand the MD copy to the physician, not wanting to disturb her as she dictates the "H and P," the procedure notes, and then electronically enters the admit physician orders. In ED Room 1, the RN is finishing entering the admit assessment, his progress notes, and creating a critical pathway but looks up to say, "Thanks for your help," as you drop off the PCR.

When all is said, done, and written, what is documented in the narrative either makes or breaks the PCR. Without an effective narrative, documentation is incomplete. The Focused EMS Event Summary allows the EMS professional to focus on the critical elements of narrative documentation, thus reinforcing the appropriateness of the EMS professional's management of the EMS event.

CHAPTER REVIEW

Review Questions

Please refer to Answers to Chapter Review Questions at the back of this book.

1. List the functions of narrative documentation.

2. List the fundamental principles of narrative documentation.

3. List the essential elements of the Focused EMS Event Summary.

4. List the essential documentation elements of the EMS Review of Systems.

5. Describe the purpose of the procedure note. Why is it important?

Critical Thinking

Please refer to Answers to Critical Thinking Discussion Exercises at the back of this book.

1. Evaluate each of the common narrative formats: SOAPIER, CHARTE, Head-to-Toe, and Review of Systems. What are the strengths and weaknesses of each of these formats? What information is captured by the Focused EMS Event Summary that the other formats leave out?

2. Why is the use of a consistent narrative format important?

3. The statement "EMS professionals do not diagnose" has been accepted since the birth of EMS. Is this true for today's EMS practice? Is it harmful to the advancement of the EMS profession to discount the diagnostic capabilities of the EMS professional?

4. What is the difference between the EMS Diagnosis and the medical diagnosis?

Action Plan

1. Apply the principles of this chapter to narrative documentation.

2. If your EMS system's PCR does not contain appropriate documentation fields that capture a complete and thorough physical examination, consider documenting the assessment utilizing the EMS Review of Systems format.

3. Begin utilizing procedure notes to document the essential information related to treatment interventions.

4. If your system does not mandate the use of a narrative format, consider implementing the Focused EMS Event Summary in your own practice.

Practice Exercises

Evaluate the following narrative statements.

1. *IV NS started @ TKO.*
   - What key information is missing?

   - What additional information would you document?

   - How could this intervention note be rewritten?

2. *Patient intubated and bagged with BVM.*
   - What key information is missing?

   - What additional information would you document?

   - How could this intervention note be rewritten?

3. *Fall, with pain to hip.*
   - What key information is missing?

   - What additional information would you document?

   - How could this narrative statement be rewritten?

CHAPTER 8

Medical Necessity

Key Ideas

Upon completion of this chapter, you should know that:

- The EMS professional is responsible for accurate documentation of medical necessity information.

- EMS services are considered medically necessary when coverage requirements have been met.

- Reimbursement decisions are documentation-based decisions.

- There are seven documentation fields that are critical to medical necessity.

- The Medical Necessity Statement summarizes the essential elements for evaluation of medical necessity.

- Documentation of medical necessity for nonemergencies requires additional information and takes additional effort.

- The Physician's Certification Statement (PCS) form provides physician certification as to the medical necessity for nonemergency EMS services.

- The Advance Beneficiary Notice (ABN) advises Medicare beneficiaries of the possibility a nonemergency transport may not be covered by Medicare.

- Documentation of medical necessity for air transport requires additional information and meticulous documentation to support its use.

- EMS professionals must embrace their ethical responsibilities in PCR documentation of medical necessity.

CASE Study _____

It is that time of the year again, the annual mandatory in-service training. You may not look forward to the skill competency stations, HAZMAT, "Right to Know," and the protocol test, but it does give you the chance to see peers you run into only at the in-service. The annual visit from the billing department supervisor is another thing altogether. Year after year it seems as though it's the "same song, same band" when it comes to the documentation lecture.

As you take your usual seat in the training room, the billing supervisor takes the podium, with stacks of "bad example" PCRs in hand. You can't help but think, "Here it comes." For the next hour, his voice racked with frustration, he reviews the same information now nine years in a row for you: "Medicare is cutting back"; "Give us more information"; "Don't forget ZIP codes"; "Just the diagnosis is not enough"; and, your favorite, "You're not telling us enough about medical necessity. Tell us why the ambulance was needed."

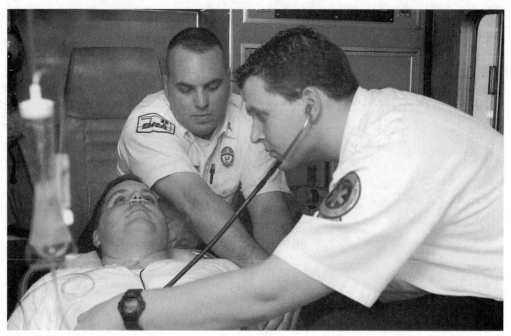

FIGURE 8-1
(Courtesy EMSA, Tulsa, OK)

An hour later, you step out for your favorite caffeinated beverage, and the usual commentary has already started outside the training room:

- "With so many important issues in EMS right now, why is this medical necessity stuff an international crisis?"
- "They have 10 full-time people. Why can't they get the information? It's not like we have the time to do 'wallet biopsies' on every patient."
- "I don't buy this medical necessity stuff and I'm not doing it. I think it's fraud."

Going back in for HAZMAT, you wonder why the same material is presented year after year and progress never seems to be made. Obviously, there is something to medical necessity, otherwise the subject wouldn't keep coming up. Everyone seems frustrated about medical necessity, from the billing department to your peers on the street. As you take your seat you wonder, What is it that we're missing about medical necessity?

Questions

Please refer to Answers to Case Study Questions at the back of this book.
1. Why is medical necessity often frustrating and controversial?

2. Is medical necessity documentation important to EMS? Why or why not?

3. Is determining and evaluating medical necessity your job, or is it the job of the billing department staff?

Introduction

Frustration and controversy seem to dominate many discussions of documentation and medical necessity. Perhaps this is due to conflicting directions that many EMS professionals have received regarding medical necessity. Perhaps discussion of medical necessity and documentation generates controversy because we do not understand:

- What medical necessity is.
- The purpose of medical necessity.
- The role of PCR documentation in determining medical necessity.
- The fundamental rules governing medical necessity.
- The EMS professional's responsibility with respect to documenting medical necessity.

Traditionally, EMS professionals have stayed clear of the subject of medical necessity. The standard approach has been to allow the care of the patient, as documented in the PCR, to stand for itself without giving direct attention to the issue of medical necessity. Some EMS professionals have even been taught that medical necessity can't be documented, believing that doing so is akin to committing fraud. The purpose of this chapter is to end the confusion, frustration, and perhaps even the controversy regarding documentation of medical necessity.

KEY TERMS

Note: Page numbers indicate where the following key terms and definitions first appear.

medical necessity (p. 122)	Advance Beneficiary Notice (ABN) (p. 133)	Medicare Compliance Programs (p. 135)
Physician's Certification Statement (PCS) (p. 131)	tertiary care facility (p. 133)	

The Ground Rules for Medical Necessity

In order to move to proficiency in medical necessity documentation, the following rules must be applied:

1. It is not the EMS professional's responsibility to establish medical necessity. Instead his or her responsibility is to ascertain, through a thorough assessment and examination, the patient's history and condition and then to accurately document them so the EMS services can be evaluated for medical necessity.

2. Not every EMS event is medically necessary. Some EMS transports are not medically necessary and should not be reimbursed. Therefore, in this discussion, we are neither attempting to "learn" to document in such a manner as to make EMS services medically necessary nor are we attempting to make nonpayable EMS services payable by embellishing documentation in order to receive reimbursement.

3. PCR documentation must always be accurate. This discussion will not center on what to document but rather on the principles that guide the EMS professional to document medical necessity appropriately.

4. The examples given in this chapter are not documentation templates.

5. EMS services to patients must never be denied or withheld based upon medical necessity.

Medicare's View of Medical Necessity

medical necessity
A reimbursement term that defines whether or not a health care service is required versus desired.

Medical necessity is a reimbursement term that refers to whether a health care service is required (necessary) versus desired. Medical services, such as cosmetic surgery, might be desired by the patient but may not be required (medically necessary). Medicare considers EMS services (emergency and nonemergency services) to be payable when the Medicare program requirements for coverage have been met. Medical necessity requirements for EMS services are dependent upon the following:

Beneficiary's condition must require the ambulance transportation itself and the level of service provided in order for the billed service to be considered medically necessary.[1]

The coverage requirements that define medical necessity are grounded in federal statute. In order for EMS services to be payable, they must meet Medicare's program requirements as set forth in the Social Security Act (§) 1861 (s) (7) and other regulations.

Medicare will pay for EMS services, emergency and nonemergency, when beneficiary's medical condition at the time of transport is such that other means of transportation, such as taxi, private care, wheelchair van, or other type of vehicle, is contraindicated (i.e., would endanger the beneficiary's medical condition).[2]

Emergency Coverage Guidelines

Medicare considers a payable EMS emergency transport as follows:

An emergency transport is one provided after the sudden onset of a medical condition that manifests itself with acute symptoms of such severity that the absence of immediate medical attention could reasonably be expected to:

- *Place the patient's health in serious jeopardy,*
- *Result in serious impairment of bodily functions, or*
- *Result in serious dysfunction of any bodily organ.*

Symptoms or conditions that may warrant an emergency ambulance transport include, but are not limited to:

- *Severe pain or hemorrhage,*
- *Unconsciousness or shock,*
- *Injuries requiring immobilization of the patient,*
- *Patient needs to be restrained to keep from hurting himself or others,*
- *Patient requires oxygen or other skilled medical treatment during transportation, and*
- *Suspicion that the patient is experiencing a stroke or myocardial infarction[3]*

Medicare coverage requirements for EMS emergencies are based upon the emergent nature of the patient's symptoms and the harm that could come to the patient if EMS services were not used.

[1]Wisconsin Physicians Service (WPS), *National Coverage Provision Ambulance Services* (2007). Retrieved May 3, 2007, from www.wpsic.com/medicare

[2]Department of Health and Human Services, Office of Inspector General, *Medicare Payments for Ambulance Transports* (January 2006), p. 3.

[3]Ibid., p. 9.

Nonemergency Coverage Guidelines

Medicare considers a payable nonemergency transport as follows:

> *Nonemergency transportation by ambulance is appropriate when a patient is bed-confined AND his/her condition is such that other methods of transportation are contraindicated: OR if the patient's condition, regardless of bed-confinement, is such that transportation by ambulance is medically required (e.g., the patient is combative and a danger to himself or others). While bed-confinement is an important factor to determine the appropriateness of nonemergency ambulance transports, bed-confinement alone is neither sufficient nor necessary to determine the coverage for Medicare's ambulance benefits.*
>
> *To be considered bed-confined, the patient must meet all three of the following criteria:*

- *Be unable to get up from bed without assistance,*
- *Be unable to ambulate, and*
- *Be unable to sit in a chair or wheelchair[4]*

Medicare coverage requirements for nonemergencies are based upon the patient's bed-confinement status and other factors that would pose a danger to the patient if ambulance transport were not used.

Regardless of Medicare's definitions of *emergency* and *nonemergency*, the EMS professional must consider every EMS encounter to be necessary, because someone believed it necessary to call for EMS services. An EMS encounter is medically necessary for *reimbursement* if the coverage requirements are met in that the patient's health could be at risk if the service weren't provided. The EMS professional needs to be familiar with coverage requirements governing medical necessity in order to capture the essential medical necessity information from the EMS encounter.

Documentation and Medical Necessity

Medical necessity cannot be ascertained apart from clinical documentation. Although other factors weigh into medical necessity, it is clinical documentation that will be used to evaluate the EMS services against coverage and level of service requirements. Therefore, PCR documentation must specifically address medical necessity.

It is critical to understand that the EMS professional does not determine whether EMS care is medically necessary. This is the role of the billing staff as its members evaluate PCR documentation in light of coverage requirements. Yet the billing department depends on the EMS professional to obtain the medical history, conduct the physical assessment, provide treatment, and document this information. EMS providers must demonstrate through PCR documentation that EMS services billed to Medicare (and other federal health care programs) as medically necessary services are truly medically necessary.

> *Medicare contractors will rely on medical record documentation to justify coverage, not simply HCPCS code or the condition code by themselves.[5]*

Because reimbursement decisions are based on documentation, the manner in which medical necessity is documented determines how EMS services will be billed to Medicare and insurance companies. If medical necessity is appropriately addressed in the PCR, the claim will be billed with accurate condition and level of service codes, resulting in an accurate payment decision.

ON TARGET Documentation of medical necessity is often erroneously thought to be only about obtaining payment for the EMS services. Documentation of medical necessity is ultimately concerned with obtaining the correct payment for the EMS services.

[4]Ibid.

[5]Centers for Medicare and Medicaid Services, *Medicare Claims Processing Manual* (2007). Retrieved May 1, 2007, from www.cms.gov/manuals

Medicare does not randomly deny claims for EMS services. Medicare denials for EMS services usually occur after review of PCR documentation because the claim is for a noncovered service or, most commonly, because "documentation does not support medical necessity." With this denial, Medicare is not saying that the EMS services were not necessary, but that the *documentation* does not support their necessity. The EMS care and transport may have been absolutely necessary; but if PCR documentation fails to address medical necessity coverage requirements accurately, the claim can (and should) be denied.

The Essential Documentation Elements of Medical Necessity

There are seven documentation fields of the PCR that are critical to medical necessity (see Figure 8-2). When clinical documentation is reviewed for medical necessity, the entire document will make a unified statement as to the presence or absence of the medical necessity of the EMS services. Beginning with the diagnosis, followed by the treatment and interventions, a clear portrait of the EMS event will enable the reviewer to ascertain medical necessity properly.

Demographics

First, demographic information is essential because it allows the claim to be billed to Medicare, Medicaid, and other payment sources. Without the patient's demographic information, the claim cannot be billed.

The EMS Diagnosis

The EMS Diagnosis is perhaps the most critical medical necessity component. As the medical conclusion (diagnosis) made by the EMS professional, the EMS Diagnosis will correlate to the condition code in the billing process. Although diagnosis alone does not establish medical necessity, diagnosis does direct the assessment and interventions that will ultimately be used to determine coverage and level of service. The following principles for appropriately documenting the EMS Diagnosis should be utilized when documenting medical necessity:

1. The EMS Diagnosis is the medical conclusion made by the EMS professional that reflects the patient's chief complaint and the associated illness or injury.
2. A medical procedure is not a diagnosis. Consider the following examples:

Procedure	Diagnosis
"X-Rays"	"Fall, R/O Hip Fracture"
"G Tube Replacement"	"Amyotrophic Lateral Sclerosis"

In these examples, the patient may have had X-rays, as a result of the fall and possible hip fracture, or the patient may have had a G tube replacement as a result of paralysis, in turn as a result of an end-stage neurological disease. As you can see, the procedure is not the diagnosis, but a result of the diagnosis.

Affirmation
Medical Necessity of Emergency Medical Service:
EMS Diagnosis: Condition Code:
Past Medical History Contributing to This EMS Incident:
Bed-Confinement Status: At the time of the transport was the patient: 1. Unable to get up from bed without assistance?   Y   N 2. Unable to ambulate; and   Y   N 3. Unable to sit in a chair or wheelchair?   Y   N Reason for Above Bed-Confinement: Contributing Diagnosis:
Inter-Facility Transports: Service(s) Not Available at Sending Facility: Immediate Transport Needed: Reason: Time Requirement: Equipment Required:
Medical Necessity for Transport to a Medical Facility of Greater Distance: Facility Bypassed: Services or MD Unavailable: Return to Facility Post Procedure: Procedure: _____   Date: _____ Patient/Family Preference Only: Y   N Other:
Sending MD: **Receiving MD:**
PCS Attached: Y N   Not Applicable **ABN Signed:**   Y N   Not Applicable/Emergency Patient
Transport/Loaded Miles:
Medical Necessity Not Established: Reason:
Medical Necessity Statement:

FIGURE 8-2
EMS Documentation PCR—Medical Necessity

3. The EMS Diagnosis must be as specific as possible. Consider the following example:

Incorrect		**Correct**
"Fall"	versus	"R/O Right Hip Fracture: Severe right hip pain post fall from stepladder"

4. The EMS Diagnosis, when written for medical necessity purposes, should be in a consistent format that includes the EMS Diagnosis and the patient's chief complaint, as demonstrated in the preceding correct example. See Table 8-1 for examples of properly written EMS Diagnoses.

Past Medical History

The patient's past medical history is a key component in determining medical necessity. PCR documentation must reference the past medical history that is pertinent to the EMS encounter. This is absolutely essential for nonemergency or interfacility transports in which the

ON TARGET Diagnosis or condition code by itself does not determine, establish, or provide support for medical necessity.

Table 8-1	EMS Diagnosis Examples
Marginal	**Descriptive**
Dementia	Alzheimer's disease: Confusion and disorientation
Stroke	Cerebral vascular accident with left hemiplegia and aphasia
Cancer	Liver cancer with abdominal distention due to ascites
Fall	Cervical spine pain secondary to 10-foot fall

patient's ambulation status must be evaluated. Key elements of past medical history critical to medical necessity documentation include:

- How does the past medical history impact this EMS event?
- How does the past medical history affect the patient's mobility?
- How does the past medical history impact the patient's safety?

Obtaining a patient's medical history is both a requirement for all levels of EMS licensure and an essential element in providing documentation for consideration of medical necessity. The following principles should be utilized when documenting past medical history:

1. The most fundamental principle is simple: The EMS professional must obtain the medical history for every EMS patient encounter (emergency and nonemergency). Take the time to obtain the information.

2. Know where to find the patient's medical history. Because patients do not always know their entire medical histories, know where to look—family members, medications, and tools such as "vial of life"—to provide clues to relevant medical history.

3. When documenting medical necessity, identify the history relevant to the EMS encounter in question. Example: Although "diverticulitis" is relevant to an EMS Diagnosis of abdominal pain, it is not relevant to hip pain.

4. Remember to be descriptive as you document relevant past medical histories. Consider the following example:

Incorrect		**Correct**
"Cardiac History"	versus	"Myocardial Infarction with stent placement in 2005"

Assessment and Interventions

Assessment and interventions support the EMS Diagnosis and provide critical information that will be used to evaluate the EMS services for medical necessity. Key documentation elements include:

- An assessment to your level of licensure. Example: A Paramedic assessing a patient with respiratory distress would be expected to auscultate breath sounds and then appropriately document the assessment findings.
- An assessment reflecting the diagnosis. Example: Documentation for a trauma patient should reflect a head-to-toe trauma exam, appropriately noting all pertinent findings.
- Interventions corresponding to the assessment and physical examination findings and the EMS Diagnosis.

The Focused EMS Event Summary—Narrative

The narrative has value in supporting medical necessity. Typically, evaluation of medical necessity is supported by diagnosis, assessment, and treatment; however, the narrative may be used if medical necessity is in question. The Focused EMS Event Summary, as discussed in the previous chapter, is an important tool for documenting information that can support medical necessity.

Signatures

The patient's signature is a requirement for reimbursement purposes.

> *Medicare requires the signature of the beneficiary, or that of his or her representative, for both the purpose of accepting assignment and submitting a claim to Medicare. If the beneficiary is unable to sign because of a mental or physical condition, a representative payee, relative, friend, representative of the institution providing care, or a government agency providing assistance may sign on his/her behalf. A provider/supplier (or his/her employee) cannot request payment for services furnished except under circumstances fully documented to show that the beneficiary is unable to sign and that there is no other person who could sign.[6]*

Failure to obtain a signature renders an EMS claim nonbillable. Retrieving a signature after the fact is a time-consuming process that could take many hours of work. Although there are certainly instances when it is difficult or impossible to obtain a patient's signature, these represent the exception and not the rule. Simply take the time to obtain the patient's signature. In the event you are unable to obtain a signature, document the reason the signature was not obtained.

The Medical Necessity Statement

The Medical Necessity Statement is the single most important financial documentation tool in EMS documentation. It provides focus and structure for constructing a short narrative statement to appropriately summarize medical necessity information.

The Medical Necessity Statement

Traditionally, medical necessity has not been directly addressed in documentation. As reimbursement emerged as a defining issue in the 1990s, EMS systems began providing dedicated fields for the EMS professional to document why ambulance transport via stretcher was required. The Medical Necessity Statement as a tool for the EMS professional and billing staff serves the following purposes:

- Brings direct focus to evaluation of the medical necessity of the EMS services in question.
- Provides structure for documenting medical necessity.
- Assists Medicare and other payers in making an informed payment decision.
- Provides for consistency in documentation of medical necessity.

Table 8-2 compares traditional documentation of medical necessity with examples of Medical Necessity Statements.

[6]Centers for Medicare and Medicaid Services, *Medicare Benefit Policy Manual* (2004). Retrieved May 1, 2007, from www.cms.gov/manuals

Table 8-2 A Comparison of Traditional and Medical Necessity Statements

Traditional	Medical Necessity Statement Examples
MVC—R/O multiple system trauma.	*Chest Trauma/Probable Pneumothorax:* Emergent transport via ALS ambulance required from scene of high-speed motor vehicle collision. Patient requires spinal immobilization, cardiac monitoring, and IV fluid replacement therapy due to possible life-threatening condition.
Transported to ED for evaluation of right hip pain post fall.	*Possible hip fracture secondary to fall:* Patient requires ambulance transport due to need for spinal immobilization and immobilization of lower extremities (to protect from further injury) requiring patient to be transported supine on long spine board.
Transport due to chest pain. R/O MI.	*R/O Anterior Myocardial Infarction w/Chest Pain:* Emergent transport and ALS interventions (12-Lead EKG monitoring, IV infusions of dopamine and nitroglycerin) required due to possible life-threatening condition. Transported to closest medical facility with interventional cardiac services.
Stretcher transport back to nursing home after G tube replacement.	*Stroke/CVA:* Patient returned to nursing home via ambulance after treatment at Municipal Hospital endoscopy for G tube replacement. Transport via ambulance stretcher required due to bed-confinement secondary to cerebral vascular accident: Patient maintains a fetal position at all times due to contractures of all four extremities. Requires lifting to/from bed, unable to turn in bed without assistance.

Writing the Medical Necessity Statement

The Medical Necessity Statement summarizes the essential elements for evaluation of medical necessity. It is comprised of the EMS Diagnosis, a statement summarizing the medical necessity (as appropriate) of the EMS services, followed by supporting assessment findings,

Table 8-3 Constructing the Medical Necessity Statement

EMS Diagnosis	Medical Necessity	Supporting Information
R/O Hip Fracture— Severe right hip pain secondary to fall from standing position.	EMS services required due to severe pain to right hip, with shortening and external rotation indicative of possible hip fracture.	Patient required head-to-toe spinal immobilization, stabilization of right hip to prevent further injury, and IV pain management.
Laceration—Partial thickness laceration to left thumb.	EMS services requested due to no other means of transport to hospital.	Patient ambulated to stretcher.

interventions, and any special considerations such as the need for specialized services, behavioral states, and the potential for deterioration. Table 8-3 provides examples of constructing the Medical Necessity Statement. The format and structure of the Medical Necessity Statement will focus the reviewer on the presence or absence of medical necessity so appropriate reimbursement decisions can be made.

Medical Necessity Considerations for Emergencies

Medicare makes no guarantee of coverage for emergencies based upon the emergency response alone. The patient's condition must be such that transport by other means would present a danger to the patient's health and would thus be contraindicated. Therefore, medical necessity for emergencies should never be assumed. When documenting for evaluation of medical necessity, answer the following questions:

- What contraindicates transport by other means?
- What places the patient's health at risk?
- What interventions are required?
- What makes the patient's symptoms emergent?

Medical Necessity Considerations for Nonemergency or Interfacility Transports

Medical necessity documentation for nonemergency or interfacility transports is commonly underdocumented and can be the most challenging for two reasons. First, nonemergency transports have the greatest risk for documentation inaccuracies because medical necessity may not be initially obvious. Assumptions or conclusions regarding medical necessity cannot be made until the patient's condition and medical history have been thoroughly evaluated. Medical necessity documentation must accurately record the patient's condition at the time of ambulance transport and the treatment provided.

Second, documentation of medical necessity for nonemergencies requires additional information and takes additional effort. Key information from the patient's past medical history is required relating to ambulation and bed-confinement status as well as to the relationship between the past medical history and the current transport. This information will come from three sources:

- The patient (or family members).
- The primary caregiver, usually the patient's primary RN at both the sending and receiving facilities.
- The medical records from sending and receiving facilities: the "H and P" (Health and Physical) and other medical history documentation from the sending facility.

Although the nonemergency is not a "load and go," it is frequently approached as such in PCR documentation. The most common cause of failure to provide support for evaluation of medical necessity in documentation is the failure of EMS providers to obtain the necessary information. Medical necessity documentation for nonemergency transports cannot be supported apart from a full assessment of the patient's condition at the time of transport and the corresponding past medical history. Medical necessity for nonemergency and interfacility transports is supported if:

1. The patient is bed-confined. Medicare considers a patient bed-confined if all of the following apply: the patient is unable to get up from bed without assistance and is unable to

ambulate and is unable to sit in a chair or wheelchair. It is the responsibility of the EMS professional to assess the patient's bed-confinement status thoroughly.

Nonemergency or interfacility transports require a thorough assessment of the patient's ability to move independently. The EMS professional must evaluate all of the following:

Ambulation Status

- Does the patient walk independently?
- What assistance is needed for the patient to ambulate (additional person, walker, or cane)?
- Does the patient sit independently in a chair or wheelchair?
- What is the patient's normal resting position?
- Example: "Patient is bed-confined at nursing home unless moved via mechanical lift by nursing home staff."
- Example: "Patient ambulates only to/from his wheelchair."

Transfer Status

- Does the patient move independently?
- How does the patient transfer from bed to stretcher?
- How many assistants are used to move the patient, and how is the patient typically moved?
- Example: "Patient does not move from nursing home bed without assistance of 4 staff members and is unable to sit in chair or wheelchair."

What Type of Assist Devices Does the Patient Use?

- Walker, cane, or lift device.
- Wheelchair (type of wheelchair—conventional, bariatric, or Geri Chair).
- Posey or other device to secure patient in wheelchair.
- Example: "Patient is wheelchair-confined but requires physician-ordered restraint device to prevent falls from wheelchair when not in bed. Transport via stretcher as patient requires constant safety monitoring secondary to confusion related to Alzheimer's."

Documenting bed-confinement requires investigation, critical thinking, and patience. Figure 8-3 provides a correct example of a Medical Necessity Statement for a bed-confined patient.

In the event that a closer destination hospital is bypassed, you must document the reason why. What is the reason for medical treatment at the sending and receiving facilities?

- Was there a change in the patient's status that now requires transport by ambulance?
- What was the diagnosis at the sending facility (the reason the patient received care at the sending facility)?

Medical Necessity Statement
Amyotrophic Lateral Sclerosis (ALS): EMS services required post-evaluation at emergency department for respiratory distress. Stretcher required due to full-body paralysis rendering patient bed-confined: required four person lift to/from bed, unable to turn without 2 person assist. Continuous oxygen therapy and frequent oral suctioning required to protect patient's airway and respiratory status.

FIGURE 8-3
Documentation of Bed-Confinement

Medical Necessity Statement
Humorus Fracture: Patient transferred from Municipal Medical Center to Regional Medical Center for services unavailable at sending facility (orthopedic treatment of humorus fracture). Emergency transport by stretcher required due to instability of fracture requiring patient to maintain semi-Fowler position and need for monitoring as patient medicated with 10 mg of morphine prior to departure from Municipal Medical Center.

Medical Necessity Statement
Acute Cerebral Vascular Accident (CVA): Emergent ALS transfer from St. Mary's Hospital within one hour of onset of symptoms (aphasia and left hemiplegia) to University Medical Center for interventional neurological services unavailable at sending facility. Regional Medical Center, the closest facility, bypassed due to interventional neuro-radiologist not available at time of transport.

Medical Necessity Statement
Acute Renal Failure: Emergent ALS transfer from Memorial Hospital to University Medical Center for transplant services unavailable at sending facility. Metro Medical Center bypassed to return patient to facility and physician where kidney transplant performed. Patient required oxygen, cardiac monitor, and meticulous monitoring of IV infusion rates en route.

FIGURE 8-4
Medical Necessity Statement Examples—Nonemergency/Interfacility Transports

- What treatment was not available at the sending facility?
- What was the reason for the treatment received at the sending facility?
- What is the planned treatment at the receiving facility?
- What treatment is required en route to the receiving facility?

Medical necessity cannot be supported without information regarding the sending and receiving medical facilities. Figure 8-4 provides correct examples of Medical Necessity Statements for interfacility transports.

2. Although bed-confinement is not the most crucial coverage requirement for the support of medical necessity for nonemergency or interfacility transports, medical necessity can also be established by certain factors that would render transport by other means contraindicated. Examples include:

- Patients requiring constant visual monitoring due to confusion or other impairment that would place the patient at risk by another means of transport.
- Patients requiring the use of restraint devices during transport.
- Patients requiring continuous visual monitoring due to a medical or behavioral condition.

Physicians generally have a good understanding of when stretcher transport is appropriate. Therefore, when called for a nonemergency transport, the EMS professional should investigate the patient's history to ascertain the reason stretcher transport was ordered. If medical necessity is not obvious, look deeper into the patient's medical history. Don't assume there is no support for medical necessity until a thorough assessment has been completed.

Physician's Certification Statement (PCS)
The PCS form is a CMS-mandated form that requires physicians to certify medical necessity for the nonemergency patient.

The Physician's Certification Statement (PCS)

Physician's certification is a Medicare requirement for nonemergency transports. The **Physician's Certification Statement (PCS)** form provides a physician's certification as to the medical necessity for nonemergency EMS service. The PCS form requires a physician to specify that the patient's condition renders transport by ambulance medically necessary. The following nonemergency transports require a PCS form:

- Nonemergency, scheduled, repetitive ambulance services.
- Nonemergency unscheduled or scheduled ambulance services on a nonrepetitive basis.

Physician's Certification Statement (PCS)
Nonemergency Ambulance Transportation – MEMS
Dispatch: 301-1407, Office: 301-1400, FAX: 301-1436

MEMS Run Number: _____

Patient's Name: _____ Date of Transport: _____

Social Security: _____ Transport From: _____

Date of Birth: _____ Transport To: _____

Medicare I.D.: _____ Private Insurance: _____

Medicaid I.D.: _____ Policy / Group: _____

In my professional medical opinion, this patient requires transport by ambulance and should not be transported by other means. The patient's condition is such that transportation and observation by medically trained personnel is required.

Check all appropriate boxes for the above named patient:

- ☐ Unable to get up from bed without assistance; ambulate; or sit in a chair, including a wheelchair, due to other conditions indicated in the narrative below.
- ☐ Dementia, Late stage Alzheimer's, Severe Altered Mental Status, decreased level of consciousness.
- ☐ Frail, debilitated, extreme muscle atrophy, risk of falling out of wheelchair while in motion.
- ☐ Requires oxygen. Liters per minute_____.
- ☐ Requires airway monitoring or suctioning during transport.
- ☐ Requires cardiac EKG monitoring during transport.
- ☐ IV Maintenance required during transport.
- ☐ Comatose and requires trained personnel to monitor condition during transport.
- ☐ Seizure prone and requires trained personnel to monitor condition during transport.
- ☐ Medicated and needs trained personnel to monitor condition during transport.
- ☐ Suffers from paralysis or contractures. _____Lower Extremities _____Fetal
- ☐ Danger to self and others, requires restraint. _____verbal _____chemical _____physical _____flight risk
- ☐ Has decubitus ulcers and requires wound precautions. _____buttocks _____sacral _____ back_____ hip
- ☐ Requires isolation precautions (VRE, MRSA, etc.) or other special handling during transport.

Narrative _____
_____

I certify that the above information is true and correct based on my evaluation of this patient, to the best of my knowledge and professional training. I understand this information will be used by the Health Care Financing Administration (HCFA) to support the determination of medical necessity for ambulance services.

_____   _____
Printed name of physician   Date

_____
Signature of M.D., P.A., R.N., R.N.P., C.N.S., or Discharge Planner (see PCS Instructions)

> Definition of Bed Confined: "Medicare covers ambulance services only if they are furnished to a beneficiary whose medical conditions is such that other means of transportation would be contraindicated. For non-emergency ambulance transport, the following criteria must be met to ensure that ambulance transportation is medically necessary: (1) The beneficiary is unable to get up from bed without assistance, (2) The beneficiary is unable to ambulate, and (3) The beneficiary is unable to sit in a chair or wheelchair." (Medicare Provider Manual)

Complete if appropriate

Hospital Discharge to Out of Town Residence: Treatment received _____

Hospital to Hospital Transfer: Elevated Care Needed _____

FIGURE 8-5

PCS Form (Courtesy of Metropolitan Emergency Medical Services [MEMS], Little Rock, AR)

Note: This PCS form is provided for educational purposes only. It must not be used to direct medical necessity documentation or as a template for PCS form creation.

A physician's certification is *not* required for a beneficiary residing at home or in a facility who is not under the direct care of a physician. However, the presence of physician's certification alone does not establish the medical necessity of the transport. Figure 8-5 provides an example of a PCS form.

The Advance Beneficiary Notice (ABN)

Advance Beneficiary
Notice (ABN)
The ABN form is a
CMS-mandated form
that advises patients
(and family members)
that Medicare may deny
claims for services that
may not be deemed reasonable and necessary.

The **Advance Beneficiary Notice (ABN)** advises Medicare beneficiaries of the possibility that a nonemergency transport may not be covered by Medicare. When CMS introduced the Advance Beneficiary Notice (ABN), it gave EMS an effective customer service tool that assists both the EMS professional and the patient. The ABN is an easy way for the EMS professional to inform patients and their families, in advance, that a service may not be covered by Medicare because nothing is more bothersome than getting a bill for services that one thought would be covered. One note of caution: Advance Beneficiary Notices must not be given to emergency patients. Follow your organization's guidelines in the use of the ABN.

Medical Necessity Considerations for Air Transports

The introduction of an air ambulance into an EMS scene is very visually exciting. However, flying at an altitude of 2,000 feet, the bar of air ambulance documentation is raised and it is easy to understand why. With an average price tag of well over $15,000 for air transport, this valuable, but often overused, medical resource must be grounded in medical necessity. Medicare carriers may elect to review every air ambulance claim prior to payment for medical necessity.

Medical necessity for the use of an air ambulance is supported only by the presence of a life-threatening condition requiring an expedited transport to a **tertiary care facility.**

tertiary care facility:
A health care facility
that is able to provide
the highest level of
specialized diagnostic
and treatment service.

Medical appropriateness is only established when the beneficiary's condition is such that the time needed to transport a beneficiary by ground, or the instability of transportation by ground, poses a threat to the beneficiary's survival or seriously endangers the beneficiary's health.[7]

Medicare provides the following examples of conditions that could justify use of air transport:

- Intracranial bleeding requiring neurosurgical intervention.
- Cardiogenic shock.
- Burns requiring treatment in a burn center.
- Conditions requiring treatment in a hyperbaric oxygen unit.
- Multiple severe injuries.
- Life-threatening trauma.

These examples illustrate the *type* of air transport that could be medically necessary. Documentation of medical necessity for air transport can be challenging. These patients are generally

[7]Ibid.

critical, and there is not much time for lengthy investigations of the patient's medical history. Documentation considerations differ for the scene flight as opposed to the interfacility transport.

Considerations for Scene Flights

When documenting scene flight responses, it is important for the following to be ascertained and documented:

- Requesting agency (Emergency Medical Responders, law enforcement, or an EMS provider).
- Level of licensure of the requesting EMS provider.
- Specifics of the mechanism of injury and how they relate to medical necessity for air versus ground transport.
- Delays in transport by ground such as the need for prolonged extrication. How long did the extraction take and were air services on scene during the extrication?
- Why was ground transport not used? How would transport by ground have been harmful to the patient?

Considerations for Interfacility Air Transports

When documenting interfacility transports by air ambulance, the following must be ascertained and documented:

- A description of how ground transport would endanger the patient's health.
- A description of the medical services that are not available at the sending facility. Example: "Transport by air as neuro ICU is not available at sending facility."
- If transport by air is requested due to the need for specialized services such as interventional cardiac or neurological services unavailable at the sending facility, this must clearly be documented. If the physician is requesting, due to medical necessity, that the patient arrive at the tertiary care facility within a certain window of time, this should be clearly and accurately documented. Example: "This 28-week high-risk OB patient is being transported by air as physician order states maximum transport time of 30 minutes to receiving hospital NICU unit. Ground transport time = 70 minutes."
- If the transport is solely at the request of the patient or family, this must be documented in the PCR.
- If a closer facility that is able to provide the same level of care is bypassed, you must document the reason for the additional air miles to the destination facility. Example: a patient returning to the facility where a major surgical procedure was done, such as an organ transplant.
- Estimated transport time by ground should always be documented in the Medical Necessity Statement.

Transport time is a major factor in determining appropriate use of air ambulance. Medicare stipulates that if the transport time by ground ambulance is 30 to 60 minutes or more for a patient whose injury or illness requires rapid transport, then air ambulance might be appropriate. Therefore, as much as possible, documentation for all air transport patients should include a comparison of the ground versus air transport times and how it relates to the patient's condition.

Because documentation of medical necessity for air transport requires accurate, thorough, and meticulous documentation, take the time, as appropriate, to ascertain why air services were requested. Figure 8-6 provides an example of a Medical Necessity Statement for air transport.

> **Medical Necessity Statement**
> *Acute Subarachnoid Hemorrhage: Patient presented (by local EMS) unresponsive to sending hospital. Transport by air ambulance required (ground transport 60 minutes/air transport 30 minutes) due to need for immediate neurosurgical services unavailable at sending hospital. This unstable patient is intubated, ventilated with mechanical ventilator, and in anesthetized state.*

FIGURE 8-6
Medical Necessity Statement Example—Air Transport

The Ethical Considerations for Medical Necessity Documentation

It would be inappropriate to end this discussion without drawing attention to the simple, but profound, ethical responsibilities associated with medical necessity documentation.

1. You must always document medical necessity accurately and truthfully, never documenting more or less than an accurate reflection of the patient's condition.

2. Your responsibility in accurately documenting the EMS event is to provide a summary that addresses medical necessity. The EMS professional does not establish medical necessity but provides the critical information, in light of coverage requirements, so that medical necessity can be evaluated for billing and reimbursement purposes. Most EMS systems have **Medicare Compliance Programs** in place to evaluate and assist in accurate documentation of medical necessity.

3. EMS is called to respond at the discretion of the public. Callers request EMS services for a variety of reasons including simply being in need of transportation to a hospital. In situations when medical necessity is questionable (or absolutely not supported), documentation must be accurate. Whether medical necessity is supported or not, the EMS professional must always document medical necessity as it stands. In this light, the EMS professional is neutral, seeking neither to prove nor to disprove the necessity of EMS services. The EMS professional, having an understanding of the coverage requirements for medical necessity, obtains the essential information and documents in an appropriate manner. Further, the EMS professional must not have one documentation standard for emergency transports and another for nonemergency transports.

4. Certain EMS services will not (and should not) be reimbursed. When medical necessity is clearly not supported, the EMS provider documents these EMS encounters in the same manner as all transports. Figure 8-7 provides examples of Medical Necessity Statements for situations in which medical necessity is either questionable or clearly not supported.

Medicare Compliance Programs
Formal programs adopted by health care organizations that provide quality measures, education, and administrative oversight to ensure that practices mirror government standards.

> **Medical Necessity Statement**
> *Urinary Tract Infection: Transport back to residence after hospitalization for urinary tract infection. Stretcher transport requested due to weakness. Patient ambulates to and from stretcher with minimal assistance.*

> **Medical Necessity Statement**
> *Right Hip Fracture: Patient transported to Regional Medical Center at request of family. Patient under care of orthopedic services at sending facility.*

FIGURE 8-7
Medical Necessity Statement Examples—When Medical Necessity Is in Question

Summary: Return to Case Study

Your first shift after the annual in-service begins with an interfacility transport. The unstable patient is being emergently transported from the local hospital to a tertiary care facility for services unavailable at the sending hospital. As you copy demographics from the face sheet, you stop yourself short of documenting, "patient transport to University Medical Center due to chest pain." Instead, you take an extra minute to review the physician's ED notes, glance over the H and P, and ask a few extra questions of the patient's primary RN. Then you document: "Medical Necessity Statement: Acute Myocardial Infarction—Patient emergently transported to University Medical Center as interventional cardiac services are not available at Municipal Hospital. Emergency ALS transport requiring IV infusions to stabilize blood pressure. Patient is unstable and at risk for deterioration." Perhaps you're not missing it anymore.

Accurate documentation of medical necessity is challenging and will take practice to master. The EMS professional must invest the time to understand the coverage criteria that govern medical necessity and the documentation essentials for both emergency and non-emergency transports. The Medical Necessity Statement provides structure, focus, and consistency for documenting medical necessity.

CHAPTER REVIEW

Review Questions

Please refer to Answers to Chapter Review Questions at the back of this book.

1. Define medical necessity.

2. List the ground rules for this chapter's discussion of medical necessity.

3. List and describe the Medicare coverage guidelines for EMS emergencies.

4. List and describe the Medicare coverage guidelines for nonemergencies.

5. List the criteria that must be met for bed-confinement.

6. List the seven critical documentation fields for documentation of medical necessity.

7. List the purposes and advantages of the Medical Necessity Statement.

8. List sources for obtaining a patient's past medical history.

9. List and describe the elements for evaluation and documentation of ambulation status.

10. List the questions that must be asked when evaluating medical necessity for interfacility transports.

11. Describe the purpose of the Physician's Certification Statement (PCS).

12. Describe the purpose of the Advance Beneficiary Notice (ABN).

13. List and describe the EMS professional's ethical responsibilities in medical necessity documentation.

Critical Thinking

Please refer to Answers to Critical Thinking Discussion Exercises at the back of this book.

1. Does a negative attitude toward nonemergencies or interfacility transports have any effect on how medical necessity is documented?

2. What are the dangers of EMS services being paid incorrectly? Which is worse: the EMS organization not getting paid or the EMS organization being paid incorrectly?

3. When a claim for EMS services is denied, "documentation does not support medical necessity," is Medicare stating the EMS services were not necessary?

Action Plan

1. Become familiar with your EMS organization's Medicare Compliance Program.

2. Implement the Medical Necessity Statement in your EMS practice.

Practice Exercises

Evaluate the following Medical Necessity Statements.

1. *Patient transported by stretcher to TMNH post fall.*

    - What information could be missing from this statement?

    - If the information were accurate and obtained from the patient's history and examination, how could this statement be rewritten to improve evaluation of medical necessity?

2. *Patient flown to UMC. Intubated and sedated unresponsive.*

    - What information could be missing from this statement?

    - If the information were accurate and obtained from the patient's history and examination, how could this statement be rewritten to improve evaluation of medical necessity?

3. *MVC—R/O Trauma.*

    - What information could be missing from this statement?

    - If the information were accurate and obtained from the patient's history and examination, how could this statement be rewritten to improve evaluation of medical necessity?

4. *Transport to psych unit. On hold with restraints.*

    - What information could be missing from this statement?

    - If the information were accurate and obtained from the patient's history and examination, how could this statement be rewritten to improve evaluation of medical necessity?

5. *Full code—Cardiac arrest.*

    - What information could be missing from this statement?

    - If the information were accurate and obtained from the patient's history and examination, how could this statement be rewritten to improve evaluation of medical necessity?

6. *Fall—Transport to ED for evaluation.*
   - What information could be missing from this statement?

   - If the information were accurate and obtained from the patient's history and examination, how could this statement be rewritten to improve evaluation of medical necessity?

7. *Level 1 STEMI—Transport to cath lab.*
   - What information could be missing from this statement?

   - If the information were accurate and obtained from the patient's history and examination, how could this statement be rewritten to improve evaluation of medical necessity?

CHAPTER 9

Putting It All Together

Key Ideas

Upon completion of this chapter, you should know that:

- The EMS Documentation Process brings together the key concepts of EMS documentation presented in this text.

- Skill in time management enables the EMS professional to consistently deliver quality patient care that is reflected in documentation, regardless of how busy the shift is.

- There is a relationship between the quality of EMS documentation and the quality of the past medical history obtained from the patient.

- There is a relationship between the quality of EMS documentation and the quality of the assessment and examination of the patient.

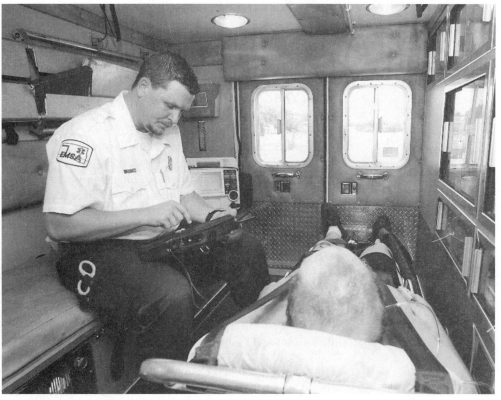

FIGURE 9-1
(Courtesy EMSA, Tulsa, OK)

It has been a tremendously long 12 hours for the 1500–0300 shift and you're extremely tired. You and your partner were in just the right place at the right time and the result: 10 calls in 12 hours. It was one of those shifts where you barely even had time to eat, but you finally were able to make a drive through before midnight. Arriving back at the station, your partner is quickly out the door and has left you with 10 PCRs to finish. It used to be much easier in the old days, you recall. All that was required was a one-page form with only name, address, a set or two of vitals, treatment, and a short narrative. These days, with so much information to enter, it easily takes 30 minutes or more per PCR.

Gathering all your cheat sheets from your 10 calls, you ask yourself: "Why does it take so long to document the demographics, scene assessment, patient history, examination, treatment, medical necessity, and patient disposition? Is there a better way?"

Questions

Please refer to Answers to Case Study Questions at the back of this book.

1. Do you see any issues in the case study with the EMS provider's approach to documentation? If so, what are the issues and what may be the solution?

2. Describe your current documentation practices. Do you complete each PCR at the end of each EMS encounter, or do you wait until the end of the shift? Do you keep up with documentation when call volume is high?

Introduction

This text has presented considerable information about EMS documentation, beginning with establishing the PCR as the professional, legal, and financial document of the EMS profession. The basics, the Five C's of Clinical Documentation, were discussed; then the EMS PCR was broken down into demographic, assessment, treatment, and affirmation components, and essential documentation elements were analyzed. The Focused EMS Event Summary was presented, which summarizes critical information from the EMS encounter. Finally, documentation for evaluation of medical necessity was discussed utilizing the Medical Necessity Statement.

This chapter introduces the EMS Documentation Process, a tool to assist the EMS professional in putting together the key concepts presented in this text. Other factors, however, also impact the ability to document effectively. The ultimate goal is to combine excellence in documentation with excellence in patient care. Time management and the development of history-taking, assessment, and examination skills all have a tremendous impact on the quality of PCR documentation. If time is not used well, PCR documentation might be done in a haphazard manner. If the EMS professional is deficient in history-taking, assessment, and examination skills, PCR documentation will lack critical detail. To bring it all together effectively, we must also advance these key skills. In this chapter we will identify the factors outside documentation that impact patient care and documentation excellence.

The EMS Documentation Process

The EMS Documentation Process, in bringing together the key concepts of EMS documentation that have been presented in this text, guides the documentation of the EMS professional in:

- **How to Document:** Documentation that is clear, complete, correct, consistent, and concise.
- **What to Document:** Documentation that directs the EMS professional to obtain key DATA: demographic, assessment, treatment, and affirmation data.
- **Documenting the EMS Diagnosis:** The EMS professional makes an assessment-based diagnosis that directs interventions and disposition of the EMS patient.
- **Documenting the Focused EMS Event Summary:** The Focused EMS Event Summary ties together the essential elements for summarizing the EMS event in the narrative.
- **Documenting the Medical Necessity Statement:** The EMS professional must affirm that medical necessity has been accurately and appropriately documented so that medical necessity can be evaluated for reimbursement purposes. The Medical Necessity Statement provides focus, structure, and uniformity for accurately documenting medical necessity.

The hallmark diagnostic sign of problem documentation is disagreement. All aspects of the PCR agree—EMS Diagnosis, chief complaint, assessment, and interventions.

The hallmark diagnostic sign of problem documentation is disagreement. It is vital that all aspects of the PCR agree—EMS Diagnosis, chief complaint, assessment, interventions, narrative summary, and medical necessity. The EMS Documentation Process is a tool to assist you in putting together the various aspects of PCR documentation so that they all agree, making a unified statement of the care given. Figure 9-2 summarizes the EMS Documentation Process.

EMS Documentation and Time Management

Perhaps you can relate to the case study, often finding yourself in the same situation of more calls than time for documentation. EMS personnel have virtually no control over where, when, or how they are sent. The communications center, or dispatch, controls destinies for the 8, 12, or 24 hours of shift time. Because of this lack of control, it is very easy to be undisciplined, allowing the EMS environment to control you. For this reason, many EMS professionals may find it difficult to complete documentation consistently and in a timely manner. Poor time management skills will

---

How to Document: The Five C's of Clinical Documentation–Clear, Complete, Correct, Consistent, and Concise
What to Document: EMS DATA Format – Demographic, Assessment, Treatment, and Affirmation Data
Assign: EMS Diagnosis
Summarize: Focused EMS Event Summary
Finalize: The Medical Necessity Statement

FIGURE 9-2
The EMS Documentation Process

be manifested by incomplete or delayed PCR completion. Being consistent in documentation will be put to the test during busy times. **Time management** is a developed skill enabling a person to gain more value from his or her time. Developing time management skills will improve your quality and productivity, allowing you to accomplish more with less. Consider applying the following principles of time management to your documentation practices.

1. **Begin by assessing how you manage your time while at work.** Applied to EMS, being skilled in time management enables you consistently to deliver quality patient care that is reflected in documentation, regardless of how busy the shift.

- How do you use "downtime"? Do you make it a habit to leave documentation unnecessarily incomplete between calls?
- After the transfer of care, do you take care of "essentials" or "nonessentials" before completing the PCR that will cause you to lose valuable time if the next call comes in sooner, rather than later?
- Does documentation habitually stack up on you, leaving you to finish multiple PCRs at the end the shift?
- Are you often left with a stack of random notes and EKG strips at shift's end to try and make sense of?

2. **Practice prioritization.** Setting priorities differentiates the "must do's" from the "can wait to do's" and the "no need to do's." Subtle differences in priorities often yield significant results in professional success. After the transfer of care has taken place, what will take priority: catch-up with the ED staff, coffee, and the voice mail check, or finishing documentation? The sooner the information is documented after the transfer of care, the more accurate and descriptive the PCR documentation.

3. **Complete the medical history and demographic sections of the PCR as the patient is interviewed.** Some in EMS believe this approach to be impersonal. Recording information as it is provided by the patient improves accuracy and communicates to the patient that what he or she is saying is important. It is also a better use of time to write information down on the PCR once than to write it on a note pad and then rewrite it on the PCR afterward. Copying critical information, such as patient medications, from notes can result in errors that may have serious consequences.

4. **Consider the use of a personal recording device.** These voice-activated dictation devices, which clip to a belt and are about the size of a pager, can be a valuable tool in EMS documentation. These devices can be used to:

- Capture critical event times such as medication administration.
- Dictate a procedure note for invasive interventions such as endotracheal intubation.
- Prep for the actual PCR. Key information can be dictated that flows with the pattern of the PCR, enabling the EMS provider to complete the PCR depending more on "playback" than on memory.

In essence, these devices serve as an electronic note pad and are therefore more advantageous than paper. Because all information can be erased at the end of the patient encounter, they provide better security of patient information than paper tools. However, if you use a recording device, do so under the following guidelines:

- Do not record conversations with the patient.
- Do not record the patient's personal health information, such as demographics.
- Record only treatment times or event notes.

Avoid the use of note pads or other "pre-PCR" documentation as they can be easily lost or misplaced. The presence of additional documentation could be discoverable as supplemental documentation, and if the information is not in agreement with the actual PCR, the door for misinterpretation is wide open.

The Impact of Assessment Skills in EMS Documentation

Excellence in documentation *and* patient care is also dependent upon obtaining all necessary information. One of the primary reasons that PCRs have insufficient documentation is that the information was never obtained. The EMS professional's level of proficiency in history-taking and assessment greatly impacts the quality of documentation. Our discussion of EMS documentation is incomplete until we address improving the quality of history-taking and physical assessment skills. Documentation and patient care must always be equally balanced in excellence.

Medical History

There is a relationship between the quality of EMS documentation and the quality of the past medical history obtained from the patient.

There is a relationship between the quality of EMS documentation and the quality of the past medical history obtained from the patient. The EMS Diagnosis, assessment, interventions, and medical necessity are dependent upon the ability of the EMS professional to obtain and document the patient's medical history. The Focused EMS History in Figure 9-3 lists the essential elements in obtaining and documenting a patient's medical history. Using the Focused EMS History in your practice will provide a guide for consistency and accuracy in obtaining medical histories.

Assessment and Examination

There is a relationship between the quality of EMS documentation and the quality of the assessment of the patient. The EMS Diagnosis, assessment, and interventions are all dependent upon the EMS professional's ability to perform an appropriate physical examination. There is a relationship between the quality of EMS documentation and the quality of the physical

---

1. Chief Complaint
   • Reason EMS Services Requested (state in patient's own words)
2. History of Present Problem
   • Complete Description of Onset and Symptoms
   • Sequence of Events
   • Duration
   • Previous Events ("A Typical Event")
   • Changes in This Presentation
3. Past Medical History (related to this illness)
   • General Health (patient's own words)
   • Medications
   • OTC Medications and Supplements
   • Related/Pertinent Hospitalizations/Surgeries
   • Major Illnesses
4. Personal, Family, and Social History
   • Living Situation
   • Home/Economic Conditions
   • Occupation
   • Stressors

FIGURE 9-3
The Focused EMS History

There is a relationship between the quality of EMS documentation and the quality of the physical examination of the patient.

examination of the patient. Seek continuously to improve and refine your examination skills and allow documentation to follow. The Focused EMS Physical Exam in Figure 9-4 lists the essential elements in conducting and documenting a patient's physical examination. Using the Focused EMS Physical Exam in your practice will provide a guide for consistency and accuracy in patient assessment.

Improving time management, history-taking, and assessment skills is a key to growth as an EMS professional. As you sharpen your history-taking and physical examination skills (appropriate to your level of licensure), allow your documentation to follow. The end result will be more detailed and descriptive documentation.

1. Vital Signs
   - BP/HR/RR/Oxygen Saturation and Temperature
2. General Appearance
   - Trauma/Appearance
   - Grooming
   - Emotional Status
   - Gestures/Body Language
3. Mental Status Evaluation
   - Speech (communication ability, quality/content)
   - Emotional State (depressed, angry, withdrawn)
   - Cognitive Abilities (memory, attention span)
4. HEENT
   - Trauma/Head (deformities)
   - Eyes (PERRLA)
   - Ears (drainage)
   - Nose (drainage)
5. Neurological Status
   - Trauma
   - EMS Cranial Nerve Exam
   - Motor Function/Sensory Function
6. Respiratory Status
   - Trauma/Appearance
   - Symmetry of Movement
   - Expansion/Excursion
   - Auscultation of Breath Sounds
7. Cardiovascular Status
   - Trauma
   - EKG/Rate/Rhythm
   - Heart Tones
   - Presence of JVD
   - Peripheral Edema/Peripheral Pulses
8. GI/GU Status
   - Trauma
   - Abdominal Appearance
   - Pain/Tenderness/Rebound/Referred Pain
   - Auscultation of Bowel Sounds
   - Change in Urinary Status
9. Musculoskeletal
   - Trauma/Deformities
   - Pain/Tenderness/Range of Motion

FIGURE 9-4
The Focused EMS Physical Exam

Making a Commitment

When one begins a career in EMS, a commitment is usually made to become the best EMS provider possible. For the student in an EMS program, the drive for success begins with the quest for completion of the program. In order to meet the goal, a lot of time is invested in learning the book knowledge and even more time in practicing and rehearsing clinical skills. Once the achievement of the course completion certificate, diploma, or degree is attained, the goal becomes validating the new license by becoming a proficient EMS professional. Why EMS professionals are like this is simple: it is because a commitment was made to the EMS profession, to our patients, and to ourselves to be the best we can be. Here's a challenge: transfer the same commitment and tenacity to attaining proficiency in documentation.

Summary: Return to Case Study

Two hours later, the last PCR is finished. Driving home you realize you have a few things to learn. First, you must improve your time management skills. Second, you have difficulty with PCR documentation because you often fail to get enough information from your patients in obtaining medical histories and performing complete physical examinations. At the end of this long night you realize that staying over two hours to finish documentation is really *your* problem. You need to manage your time better and to expand your history-taking and assessment skills.

The Focused EMS History and Focused EMS Physical Exam provide templates for guiding the EMS professional in becoming proficient in taking medical histories and performing physical examinations. The essential information the EMS professional obtains, combined with critical thinking, enables the EMS professional to make accurate treatment decisions, ensuring the delivery of excellent patient care. Then, using the EMS Documentation Process, this excellence can be communicated effectively in the PCR. This is putting it all together.

CHAPTER REVIEW

Review Questions

Please refer to Answers to Chapter Review Questions at the back of this book.

1. List and describe the EMS Documentation Process.

2. Discuss how time management skills impact documentation quality.

3. List the elements of the Focused EMS History.

4. List the elements of the Focused EMS Physical Exam.

Critical Thinking

Please refer to Answers to Critical Thinking Discussion Exercises at the back of this book.

1. What is the relationship between proficiency in assessment skills and proficiency in documentation?

2. Apart from documentation, list the primary clinical skills or procedures applicable to your level of licensure. How could sharpening each of these skills positively impact your documentation?

Action Plan

1. Use the EMS Documentation Process in your practice.

2. Consider purchasing a physical examination or nursing health assessment textbook. Explore ways to apply appropriate information within the boundaries of your licensure and your EMS system's medical direction.

3. Consider purchasing resources specific to skills such as auscultation of breath sounds and heart tones. Increasing your understanding and skill in any clinical area enables you to provide, and document, better patient care.

4. Analyze how you spend your time on the average EMS event. Look for ways not to cut corners but to maximize time management so that documentation is as close to "real time" as possible.

Practice Exercises—Role Play

In these role-play exercises, which are best suited for the classroom setting, putting together patient care and documentation will be practiced. For each scenario, simulate/role play (to your level of training) how you and "your partner" would manage the EMS event. Using the Focused EMS History and the Focused EMS Physical Exam, assess and manage "your patient." Then, using the EMS Documentation Process, complete a PCR.

Scenario 1: Trauma Patient from MVC—Multiple Trauma

Scenario 2: Medical Patient with Chest Pain

Scenario 3: Interfacility Transport of a Patient with a New-Onset CVA

CHAPTER 10

Patient
Refusals

Key Ideas

Upon completion of this chapter, you should know that:

- The appropriately executed patient refusal fulfills duty to act and establishes the patient's legal and medical ability to refuse EMS care.

- Duty to act extends to evaluating the patient's capacity and competency while respecting the patient's right to self-determination.

- Capacity defines a person's legal authority, or *qualifications*, to make his or her own health care decisions.

- Competency refers to a patient's *ability* to make appropriate medical decisions and can be determined only through a focused assessment of the patient's mental and neurological status.

- Self-determination is expressed by consent, the ability to say "yes" or "no" to EMS care.

- The EMS Refusal Interview assists the EMS professional by focusing attention on capacity, competency, and consent.

- Refusal documentation must communicate that the EMS professional fulfilled duty to act and appropriately established the patient's capacity, competency, and consent.

- The CASE CLOSED Refusal Narrative enables the EMS professional to capture the essentials of the refusal interview in documentation.

CASE
Study _____

You are called to the residence of an elderly female. Upon arrival, you are met at the door by a man who states, "I called you for my mother who needs to go to the emergency room to get checked out." He directs you to the living room where you note a well-dressed elderly female resting supine on her sofa. You introduce yourself as being from EMS and ask how you may assist her. She replies, "Oh, my son thinks it's an international crisis if he doesn't have something to worry about. I'm fine, actually." Your patient appears alert and oriented and in no apparent distress. As you begin questioning the patient, you ascertain she has probably had a syncopal episode and collapsed on the sofa while having a heated discussion with her son over moving to an assisted living center. She states she has no pain, feels fine, and does not want to go to the hospital. You quickly have her sign a patient refusal form and you're back in service.

FIGURE 10-1
(Courtesy Acadian Ambulance Service, Lafayette, LA)

Questions

Please refer to Answers to Case Study Questions at the back of this book.

1. How would you evaluate the manner in which this patient refusal was managed?

2. Would you have handled this patient differently? If so, how would you have managed the patient's refusal?

Introduction

A recent survey revealed that 44 percent of EMS systems allow EMS personnel to initiate patient refusals.[1] Although patient refusals are an everyday occurrence in most EMS systems around the country, no other single type of patient encounter presents more risk to EMS systems and EMS professionals. This chapter will address the complexities of the patient refusal, beginning with the legal principles that must guide the refusal process, followed by the essential elements for interviewing and documenting a patient refusal.

[1]D. Williams, "2006 JEMS 200-City Survey," *Journal of Emergency Medical Services*, 32, 2(2007), 38–54.

Legal Principles and the Patient Refusal

The appropriately executed patient refusal fulfills duty to act, establishes the patient as legally and medically able to refuse, and respects the right to self-determination.

Appropriately conducting and documenting the patient refusal is challenging due to the legal requirements that must be met to ensure the patient is legally qualified and able to understand his or her medical condition and the risks associated with refusing EMS services. For the purpose of this text, we will define a patient refusal as the right of the legally qualified and competent patient to refuse medical care after the EMS professional has educated the patient on his or her condition, the value of EMS treatment, and the risks associated with not accepting EMS services. The goal in managing refusals is to reduce the liability associated with poor outcomes by appropriately interviewing the patient and documenting the refusal. The appropriately executed patient refusal fulfills duty to act, establishes the patient as legally and medically able to refuse, and respects the right to self-determination.

The Patient Refusal and Duty to Act

duty to act
The EMS professional's legal duty to treat a patient in accordance with EMS regulations and standards of care.

The fundamental legal principle for EMS professionals is fulfilling duty to act. **Duty to act** refers to the EMS professional's legal obligation, by reason of licensure, to act in accordance with established scope of practice and standard of care. Duty to act begins when EMS is dispatched to respond and patient contact is made. Duty to act is easily met when the patient is treated and transported. The formula is usually very simple: appropriately treat, transport, and transfer care.

Stating the traditional "sign here" after hearing a patient say, "I don't want to go to the hospital," fails to meet the legal obligations of duty to act. Duty to act extends to evaluating the patient's capacity, competency, and right to self-determination. If the EMS professional fails to ascertain that the patient is able to refuse, then duty to act is breached and the EMS professional can be accused of **abandonment.**

abandonment
The termination of the EMS professional–provider relationship prior to an appropriate transfer of care.

The Patient Refusal and Capacity and Competency

capacity
Legal qualification to make a health care decision.

competency
Ability, in the medical sense, to make a health care decision.

Another fundamental legal principle related to the patient refusal is capacity and competency. A patient must be legally and medically qualified to refuse EMS services. **Capacity** is a legal term that defines a person's legal authority, or *qualifications,* to make his or her own health care decisions. Usually, a patient has the legal capacity to accept or refuse medical treatment if he or she is over 18 years of age.

Competency refers to a patient's *ability* to make appropriate medical decisions and can be determined only through a focused assessment of the patient's mental and neurological status. Competency is established when the patient meets all the following:

- Is alert and oriented to person, place, time, and event.
- Has the ability to comprehend what is being communicated to him or her.
- Has the ability to understand the benefits of receiving EMS treatment.

Capacity refers to the patient's legal qualifications whereas competency refers to the patient's medical qualifications.

- Has the ability to understand the risks of refusing EMS treatment and transport.
- Has the ability to make a decision based upon the information presented.

The competent patient not only is alert and oriented but also is able to demonstrate the ability to process information and make a decision.

The Patient Refusal and the Right to Self-Determination

self-determination
A person's right to make his or her own decisions without pressure or inappropriate influence.

Our laws provide patients the right to **self-determination.** The EMS patient, having passed the tests of capacity and competency, has the right to accept or refuse EMS care. EMS professionals walk a fine line here. On one hand, the EMS professional must not violate a patient's right to self-determination by providing EMS care against the patient's will. On the other hand, a refusal must not be obtained from a patient who has not proven capacity and competency. The critical factor in self-determination is assessment. Failure to provide care appropriately to the patient lacking capacity and competency is usually the result of failure to assess the patient thoroughly. Likewise, treating and transporting a patient against his or her will is usually the result of failure to do a thorough assessment. Self-determination is expressed by consent, the ability to say "yes" or "no" to EMS care. There are three types of consent: informed consent, implied consent, and involuntary consent.

Informed Consent

informed consent
Consent for medical treatment that is given after the patient has been fully educated.

Informed consent is the agreement to accept or refuse medical treatment. The premise of informed consent is that good decisions are based on good information. Therefore, a consent, or a refusal, is valid only after the patient has been informed (educated) regarding his or her illness or injury, proposed treatment, risks and benefits of treatment, and dangers associated with refusing treatment. In the case of a patient being conscious but having an altered level of consciousness, consent is not as simple as a "yes" or "no" answer. A "no" answer from this patient would not qualify as informed consent because he or she would not be considered competent.

Implied Consent

implied consent
Consent for medical treatment that is assumed for a patient who is unable, by reason of illness or injury, to grant informed consent.

Implied consent applies to the patient who would most likely consent to EMS but is prevented from granting informed consent as the result of injury or illness. Examples of implied consent include the unresponsive patient or the patient found in cardiac arrest.

Involuntary Consent

involuntary consent
Consent for medical treatment that is made on behalf of a patient by legal process.

Involuntary consent applies to the patient who no longer possesses the right to make his or her own decisions and give consent. Patients who give involuntary consent have court-appointed legal representatives that make decisions on their behalf, usually because they lack legal capacity to do so.

The EMS Refusal Interview

The EMS Refusal Interview assists the EMS professional in focusing attention on determining capacity, competency, and consent, which are critical to the management of the patient refusing EMS care. An interview process specific to the patient refusal is central to favorable outcomes, which reduce liability by moving the patient from refusing to accepting EMS care. If the patient persists in refusing care against medical advice, a favorable outcome is the reduction of liability by ensuring that capacity, competency, and consent have been appropriately established.

The EMS Refusal Interview promotes consistency in evaluating patients refusing EMS care and is as elementary as "A-E-I-O-U."

- *Assess*
- *Educate*
- *Inform*
- *Offer Transport*
- *Understanding*

The EMS professional, while respecting the right to self-determination, should operate under the premise that a patient will accept EMS services until the patient indicates the desire to refuse care. Unfortunately, patients can be "sold" on refusing services immediately upon EMS arrival by being asked such questions as "Do you want to go to the hospital?" or "You don't want to go to the hospital, do you?" at the beginning of the interview.

A—Assess: Assessment of Condition, Capacity, and Competency

Everything in EMS begins with assessment. The patient accepting EMS care does not need to be evaluated for competency, and consent is covered the moment the patient states, "Take me to the hospital." However, patients that either have refused EMS services or are undecided require additional avenues of assessment than those patients accepting EMS care. The EMS professional's approach to these patients must be altered and there must be a drastic change in direction.

Condition

Begin the interview by saying, "I'm with the Emergency Medical Services ambulance. How may I help you today?" in order to obtain the patient's chief complaint or condition. This is often obvious to the EMS professional, but you need to know whether it is obvious to the patient as well. If your understanding of the reason EMS was called matches the patient's (or family member's), this is valuable information providing clues into the patient's neurological, mental, or social status. Also, by utilizing this technique you will be respecting the patient's right to self-determination. Consider the following correct examples:

- *"I'm with the Emergency Medical Services ambulance. How may I help you?"*
- *"Why was the EMS ambulance called for you today?"*

Capacity and Competency

By introducing EMS to the patient, the door is opened to ascertain the patient's capacity and competency, the most critical step in the refusal interview. Determining capacity and competency does not require in-depth psychological evaluation. Seasoned EMS professionals glean this information from patients without realizing it. Few EMS professionals will leave a patient curbside, who is not alert and is unable to understand the risks of refusal, comprehend his or her illness or injury, or reach a decision. The "A-B-C-D"

> **Examination:**
>   1. Ask the patient:
>   - What is the season, year, and today's date?
>   - Where are you right now?
>   2. Tell the patient: "I will name three objects." Example: "Tree, car, house."
>   Ask the patient: "Please tell me the three objects I named for you."
>   3. Ask the patient: "Spell *world* backwards."
>   4. Ask the patient: "Repeat the three objects that I named for you a minute ago."
>   5. Ask the patient to follow a command. Example: "Place your right hand on top of your left hand."
>
> **Evaluation:**
> If the patient fails more than two out of the five evaluation steps, question competency.

FIGURE 10-2
Mental Status Examination

approach for determining capacity and competency is simple and applicable to EMS practice:

A—Alert: Is the patient alert and oriented to person, place, time, and event?

B—Behavior: Is the patient's behavior appropriate?

C—Comprehension: Is the patient alert and oriented and able to comprehend his or her condition? Does the patient understand the illness, injury, treatment options, and risks associated with refusal of EMS service? Does the patient exhibit intact memory and judgment? Is the patient's speech normal? Is the patient cooperative? Is there any evidence of drug or alcohol use? Figure 10-2 provides an example of a mental status examination that may be used to assess the patient refusing EMS services.

D—Decision: Is the patient legally qualified to make a refusal decision? Is the patient of legal age and status? In addition, if you sense unwillingness or hesitancy to make a decision, this should be a red flag. Indecision itself is often a diagnostic sign. Do not let these patients out of your sight until you are 100 percent sure that they are competent and have the capacity to refuse EMS services.

Physical Assessment

After identifying the patient's chief complaint and validating capacity and competency, the patient requires examination to the degree possible and appropriate in the EMS environment. The street corner, place of business, public area, or even the average residence is not the best place to conduct a physical assessment, especially in the case of the reluctant patient. Some patients will not allow any type of physical exam. The EMS professional must attempt to gain consent for a physical assessment by separating it from transport to the hospital. Consider the following correct examples:

- *"We're not as concerned right now about your going to the hospital as we are about making sure you're all right. May I check you out?"*
- *"May I begin by taking your vital signs?"*
- *"May I briefly examine you?"*
- *"Would you mind if we just checked to make sure your arm is OK?"*

The assessment of the patient refusing EMS services must be as complete as the assessment of a patient accepting treatment and transport.

E—Educate

The EMS professional's duty to act includes educating the patient. Because education will do more to turn around refusals than anything else, it is the most critical aspect of the informing process. For a patient to make an informed decision, the patient must be educated in the following ways:

- Educate on the condition. Avoid medical terminology that would not be understood by the layperson. Consider the following examples:

Incorrect	Correct
"You appear to be having an MI."	*"You appear to be having a heart attack."*
"It is possible you have a fracture."	*"It is possible you have broken your arm."*
"It is possible you've had a CVA."	*"It is possible you've had a stroke."*

- Educate on the results of your assessment and examination, the possible medical diagnosis, and the EMS treatment you would provide.
- Educate on the benefits of EMS services. People respond to value, specifically how a service will benefit them. The patient must see value in EMS services in order to consent to receiving EMS services. Therefore, educate the patient as to what you can do for him or her. Explain the plan of treatment the patient can also expect upon arrival at the Emergency Department.
- Educate with literature specific to the chief complaint, condition, or injury. This practice is becoming more common in EMS systems around the country, and its merits are obvious. Patient information sheets provide consistency among EMS professionals and serve as a uniform tool to educate and inform the patient thoroughly on his or her condition, associated risks, and specific instructions that should be followed. Patient information sheets also require the patient to acknowledge receipt via signature.
- Educate the patient on the limitations of EMS. The public has high expectations of EMS, and patients must understand that the EMS system does not have the diagnostic capabilities of a medical facility and the EMS provider is not a substitute for a physician.
- Educate in response to any expressed concerns from the patient. If the patient seems to be leaning toward accepting EMS services but appears hesitant, you should ascertain the reason behind the hesitancy. One of the most common areas of concern relates to the cost of EMS services. Financial concerns are very important, especially to the elderly, and should be addressed. Consider the following correct examples:
  - *"I understand your financial concerns. My first concern is for your safety."*
  - *"Most insurance plans cover EMS services. My first concern is for your safety."*

I—Inform

Next, the patient must be informed of the risks and consequences that could be associated with refusing EMS services. Be very clear that refusing EMS services could result in serious disability or death. Risk statements should be specific to the injury or illness, and be brutally blunt. Consider the following correct examples:

- *"If you are having a heart attack, you are at risk of your heart stopping and you could die."*
- *"If you have a neck injury, you are at risk for being paralyzed from the neck down."*

If the patient has a minor injury, the EMS professional must present realistic risks to the patient, such as the risk for infection as a result of a laceration. When an illness or injury is obviously non-life-threatening, stress the diagnostic limitations of EMS.

O—Offer Transport

The patient has been assessed, educated, and informed. Now offer the services, an often neglected step in providing EMS care. Offer treatment and transport and offer at least *three times*. When a patient refuses, he or she often gives a reason for refusing EMS services. Patients frequently say "no" because they require further education, and "no's" are opportunities to answer lingering questions or concerns. By offering EMS care three times, you are presenting patients the opportunity to have their concerns or fears addressed. Most importantly, you are giving yourself three chances to ensure that you haven't missed something. Alternating between you and your partner in making the offers communicates a unified caring approach to the patient.

If three offers of treatment and transport have been refused, it is important to offer an alternative. Never suggest to patients that they drive themselves to the hospital, have family members drive them to a medical facility, call another ambulance provider, or "wait and see." These alternatives may suggest that you deferred your responsibility back to the patient or a family member. The only acceptable alternatives to offer patients are:

- *"You may call EMS back at any time."*
- *"If you change your mind, you may dial 911 at any time and we will be happy to serve you."*
- *"If you develop symptoms, or if your symptoms become worse, you should dial 911 at once and another ambulance will come to you."*

U—Understand: Validate Understanding

If, despite best efforts, the patient ultimately refuses EMS services, the process must be concluded by validating understanding. This is accomplished by summarizing what you have told the patient and what the patient has told you. Validating understanding ties everything together and communicates to your patient that you take his or her refusal seriously.

To sum up, the EMS Refusal Interview is as simple as "A-E-I-O-U": assess, educate, and inform of risks; offer EMS services; and validate that the patient understands the refusal process. Figure 10-3 provides an overview of the EMS Refusal Interview.

A—Assess Patient
- **Chief complaint**
- **Capacity**
- **Competency**
- **Consent for EMS care or refusal**
- **Physical exam**

E—Educate Patient
- **Condition**
- **Exam/assessment results**
- **Benefits of EMS care**
- **Education literature**
- **Limitations of EMS**
- **Answer questions**

I—Inform Patient
- **Risk and consequences**

O—Offer EMS services
- **Offer three times—"3 Offer Rule"**

U—Understanding
- **Summarize refusal interview**
- **Validate patient understanding**

FIGURE 10-3
The EMS Refusal Interview

The EMS Refusal Documentation Process

The goal of refusal documentation is to limit liability, which can be accomplished only if documentation:

- Communicates that the EMS professional fulfilled duty to act.
- Communicates the patient's capacity and competency.
- Communicates the patient's informed consent.

As discussed in Chapter 4, negligence is the foundation for most claims against EMS providers, and the patient refusal is a prime target of plaintiff's attorneys because negligence is easier to allege when it appears nothing was done for the patient. Certain EMS providers have lost careers as a result of incomplete refusal documentation. A hastily written and incomplete refusal may not reflect the manner in which the refusal was managed, but it will be how the incident is perceived. A properly documented patient refusal transfers responsibility from you to the patient and will close the case on misinterpretation of how the patient was managed.

CASE CLOSED Patient Refusal Narrative Documentation

The CASE CLOSED Patient Refusal Narrative provides a template for capturing the essential documentation elements of the A-E-I-O-U refusal interview.

C = Condition, Capacity, and Competency

Document the reason EMS was called and the patient's chief complaint (or presenting problem), using the patient's own words. Next, document your case for the patient's competency, providing significant detail of the patient's neurological and mental status that will support the patient's ability to make an informed decision. As previously stated, it is insufficient to document "patient was alert and oriented." Instead, *describe* the patient's neurological and mental status. Essential documentation elements include:

- **General appearance:** Neat/well groomed versus inappropriate dress/poor hygiene.
- **Speech:** Clear/appropriate versus aphasia/difficulty speaking/abnormal pattern.
- **Mood and affect:** Cooperative/calm versus agitated/withdrawn/hostile.
- **Motor activity/ambulation status:** Normal/purposeful versus rapid, slow, or nonpurposeful movement.
- **Cognitive function:** Document the findings of the mental status examination.
- **Signs of alcohol or drugs.**
- **Any changes in the patient's normal behavior.**

A = Assessment

Document the results of the assessment and examination. Essential documentation elements include:

- **History of present illness or injury.**
- **Past medical history.**
- **Medications and allergies.**
- **Physical exam.**
- **Vital signs:** A minimum of two complete sets of vital signs should be recorded on a refusal: the first set, taken after you arrive on scene and begin your assessment, and a final

set before you leave the patient. This communicates an ongoing process in your approach to the patient.

Document the results of the physical exam in the same manner you would for a patient being treated and transported to the hospital. If the patient refused to allow you to perform a detailed physical exam, document what you were visually able to assess.

S = Statements Made by the Patient

Statements made by the patient are critical to a well-documented refusal, communicating the patient was engaged and participating in the refusal process. Patient statements that should be recorded include:

- Understanding of their condition.
- Understanding they are refusing Emergency Medical Services and the associated risks of refusing EMS care.
- Understanding by signing the refusal form that they are assuming responsibility for the consequences of not accepting emergency medical services.

E = Educate

Document the information you provided to the patient with respect to his or her condition and the manner in which you validated understanding.

C = Consequences

Refusal documentation must include how the patient was informed of the potential consequences of refusing EMS services and statements made by the patient indicating understanding.

L = Limitations of EMS

Refusal documentation must include how the patient was informed of the diagnostic and treatment limitations of EMS.

O = Offer Transport

Refusal documentation must include the number of times the patient was offered EMS services. In addition, the patient should sign or initial that the offer of EMS services was made multiple times.

S = Signature and Initials

The patient refusal form is a legal document that must be signed by the patient. The signatures of the following must be present on the refusal form:

- The patient or the person legally designated to make the refusal decision.
- EMS staff.
- Appropriate witnesses. Witnesses are important in the refusal process as they back up what was communicated to the patient. Any person signing a refusal as a witness must first be educated as to what is being signed. Multiple witnesses are advantageous and family members are the primary choice.

Refusal documentation must be completed while you are with the patient and never after the fact. Never ask a patient to sign a blank refusal. If a patient had a poor outcome, think of the impression that would be made if the patient's copy of the refusal were incomplete or had a blank narrative and later a "completed" copy were found after the fact in discovery. The patient must always receive a copy of the refusal document he or she signed.

If the patient refuses to sign the refusal, document the patient's exact words and obtain witness signatures attesting to the refusal to sign. In addition, a supervisor should be notified of patient refusals in which the patient is also refusing to sign the refusal document.

E = Educational Material

Document any educational materials that were given to the patient, such as patient information sheets.

D = Dial 911 Again

Refusal documentation must always include that the patient (and/or family) was specifically instructed to call EMS again immediately if there is a change in the patient's condition or if they patient simply decides to accept EMS services.

The CASE CLOSED Patient Refusal Narrative enables the EMS professional to capture the essentials of the refusal interview in documentation. Figure 10-4 provides an example of a refusal narrative using the CASE CLOSED narrative format.

C—EMS called to residence by patient's son. On arrival found 75-year-old pale and diaphoretic female supine on sofa with washcloth on forehead. Son states he called for EMS after his mother "fainted." Patient states, "I'm fine. I just felt dizzy for few minutes." Son reports he and his mother were arguing over his suggestion she move to an assisted living center.

A—Patient neatly dressed in well-kept surroundings. Patient is calm and cooperative (but seems angry with her son) with clear, well-articulated speech. Answers all questions appropriately. Patient nonambulating due to cardiac evaluation. Movement purposeful and follows commands. Mental Status: Alert and oriented to person, place, time, and event. Recall, memory, and attention span intact. Son states no change in patient's behavior, stating, "She is as stubborn as always." Patient denies previous history of syncope. Past Medical History: hypertension (medicated with atenolol). Denies allergies. Exam: (*Document results of Focused EMS Physical Exam.*)

S—Patient refused EMS services, stating: "I do not want to go to the hospital. If there is something to this, I will go later." "There is nothing that you will say that will change my mind."

E—Patient and son educated on the refusal process and refusal form prior to signing. Patient and son verbalized understanding of refusal process. Son clearly does not support mother refusing EMS care.

C—Patient informed of the risks of refusing EMS services. Patient informed: "Your symptoms may be the result of a serious condition, such as a heart attack, and you could die." Patient verbalized understanding of the risks of refusing EMS care, stating, "So what if it's my heart. At least I won't be sent off to some assisted living center."

L—Patient informed of the diagnostic limitations of EMS and that her condition required EMS care and evaluation by a physician at a hospital. Patient verbalized understanding.

O—EMS services offered to patient X3.

S—*Obtain signatures of patient and family member. (EMS staff sign in presence of the patient and family.)*

E—Patient and son given, and signed for, "Cardiac Symptoms Patient Information Sheet."

D—Patient instructed to "dial 911 again" if she changed her mind, or if a change in condition occurred. Patient stated, "You folks have been so nice. If I start feeling bad, I'll dial 911."

The acronym letters CASE CLOSED serve only as a guide in documentation and would not be included in the narrative.

FIGURE 10-4
The CASE CLOSED Refusal Narrative Example

Summary: Return to Case Study

Three months later you and your partner are pulled out of rotation to report back to base. Upon arrival you both are directed by the supervisor to the conference room where you find the director of operations, the medical director, and the quality manager awaiting your arrival. Taking a seat, the director pushes a patient refusal across the table and asks whether you remember obtaining a refusal from a woman who had suffered a syncopal episode. After reading the documentation, you are about to reply that you don't remember this patient when your partner states: "You remember the refusal where the son and the patient were arguing about the assisted living center?" The medical director makes the problem very clear. "The patient never made it to the assisted living center as she died of a myocardial infarction later that evening. The son has filed a wrongful death suit against the company, and, as a result, your employment is suspended pending further investigation."

Patient refusals present significant legal risk to EMS systems and EMS professionals. These risks can be reduced by understanding the legal principles that must direct the refusal process and applying the tools presented in this chapter. The EMS Refusal Interview provides the EMS professional with an effective guide to validate capacity, competency, and consent. The CASE CLOSED Patient Refusal Narrative format captures the essential elements that should be documented for every patient refusing EMS services.

CHAPTER REVIEW

Review Questions

Please refer to Answers to Chapter Review Questions at the back of this book.

1. Describe how the EMS professional fulfills duty to act to the patient refusing EMS services.

2. How does capacity differ from competency?

3. List the criteria that must be met for establishing competency.

4. Describe how competency is determined in the refusal interview.

5. Discuss why documenting "patient is alert and oriented" is insufficient.

6. List and describe the elements of a mental status exam.

7. List and describe the components of the EMS Refusal Interview.

8. List and describe the components of the CASE CLOSED Patient Refusal Narrative.

Critical Thinking

Please refer to Answers to Critical Thinking Discussion Exercises at the back of this book.

1. Does your EMS organization have a culture that encourages obtaining patient refusals? How does this affect the quality of patient care and documentation?

2. Does the attitude of the EMS professional toward a patient presenting with only the appearances of a minor illness or injury impact patient care and documentation?

Action Plan

1. Use the EMS Refusal Interview in your EMS practice.

2. Use the CASE CLOSED Patient Refusal Narrative format in your EMS practice.

Practice Exercises

Evaluate the following refusal narratives.

1. *The MVC.* See Figure 10-5.

EMS Documentation Refusal Narrative

EMS called for a single vehicle MVC. On arrival found patient ambulating, talking on cell phone. Refused examination. Patient states, "I'll have my wife take me to the hospital. The only thing that hurts is my right knee." Patient noted to be walking with slight limp. Patient signed refusal. Advised of risk of refusing EMS transport.

FIGURE 10-5
Refusal Narrative

- How would you evaluate this refusal documentation?

- What information is missing from this refusal?

- How could this refusal narrative be rewritten if additional information were obtained in the interview?

2. *Doing better after D50.* See Figure 10-6.

EMS Documentation Refusal Narrative

Patient refused EMS services and transport. On arrival found this 32-year-old female unresponsive, except to painful stimuli (moans). Long history of insulin dependent diabetes, well known to this EMS crew. Vital Signs per above. Administered 1 amp of D50W. Patient became alert and oriented within 5 minutes and signed refusal.

FIGURE 10-6
Refusal Narrative

- How would you evaluate this refusal documentation?

- What information is missing from this refusal?

- How could this refusal narrative be rewritten if additional information were obtained in the interview?

3. *Role-Play Exercise:* interview and document a patient refusal, using the EMS Refusal Interview (A-E-I-O-U) and the EMS Refusal Documentation Process (CASE CLOSED).

Incident
Reporting

Key Ideas

Upon completion of this chapter, you should know that:

- The incident report provides a means for the internal communication of incidents relating to patient care, safety, and internal processes for the purposes of investigation, evaluation, and correction.

- The "Five Rights of Incident Reporting" provide direction for appropriate incident report documentation.

- The Internal Communication Report is a valuable and necessary communications tool for incident reporting and is the preferred method for internal communication of incidents.

- The basic principles of documentation also apply to incident reporting. The report must be clear, complete, correct, consistent, and concise.

- Incident reporting is for internal administrative use only and must be separate from PCR documentation.

FIGURE 11-1
(Courtesy Lakes Region EMS, North Branch, MN)

CASE Study

As the premier interfacility service in the area, your service handles the majority of these transports. The latest is a response to the intensive care unit of a community hospital for an emergency transport to the cardiac care unit of a large metropolitan hospital. Upon your arrival, staff members direct you to Room 4, where you find a 65-year-old male patient who is being transferred for acute coronary syndrome. You introduce yourself to the patient and begin taking a report from the patient's primary care nurse. He states the patient presented to the Emergency Department four hours ago with substernal chest pain radiating to the left side of his neck and is stable. He "just needs to be transported to the cath lab as soon as possible." The nurse states there are no additional orders as he hands you the envelope containing the transfer orders and quickly leaves the room.

You and your partner transfer the patient to your stretcher with the help of the ICU tech. You note the patient is on 4 L of O_2 via nasal cannula and has heparin and NTG infusions. Cardiac monitor reveals a normal sinus rhythm. Vital signs are stable and the patient states he feels fine. "Whatever they did for me downstairs really worked."

Loaded up, you head out of ICU, down the hall, and into the elevator for the short ride down to the ambulance. In the elevator your patient abruptly states his chest pain is horrible as he slumps forward, unresponsive. Your reaction is swift, but in the race against disaster you note copious amounts of vomitus oozing out of the patient's mouth and nose. You instinctively reach for the suction unit and realize it is resting comfortably in the first response cabinet of the ambulance. Disaster has won. How will you document this?

Questions

Please refer to Answers to Case Study Questions at the back of this book.
1. How would you document the incident in the case study?

2. Would you use the PCR or another form or both?

Introduction

Although the case study is somewhat extreme, it is designed to alert you to the complacency often present in EMS. Every EMS encounter has associated risks, and these risks are multiplied when complacency slips into EMS practice. Incidents do happen, often despite the best effort and professionalism. If you do find yourself in one of these unfortunate situations, how do you document the event? The **incident report** is a common tool in health care and is often referred to as a "QA report" in some EMS systems. Most EMS organizations use some form of incident reporting to communicate issues ranging from equipment problems to treatment errors, to violations in policy and procedure, and so on.

Whenever an "incident" takes place, legitimate legal concerns arise and the incident must be properly documented. Something out of the norm happened. Perhaps your equipment malfunctioned on a call, a medication error occurred, or the patient was harmed in some way. In the eyes of many, if there was an incident, someone did something wrong and quality was compromised. This chapter will discuss the purpose and principles of incident report documentation.

incident report
Report of any event that is inconsistent with established guidelines, procedures, or outcomes.

The Purpose of Incident Reporting

The incident report provides a means for the internal communication of incidents relating to patient care, safety, and internal processes for the purposes of investigation, evaluation, and correction. Incident reporting is an essential tool for communicating events that could impact patient care, the EMS system, and the EMS staff. Examples include:

- Equipment issues.
- Deviations from protocol or standard operating procedures.
- Accidents and injuries.
- Patient care errors.
- Any incident that could result in a detrimental outcome for the EMS patient.
- Any incident that could result in a detrimental outcome for the EMS system.

Incident reporting allows for early identification, investigation, and appropriate corrective action. It is vitally important to communicate the appropriate information, in the correct manner, through incident reporting. Perhaps you have heard of the Five Rights of Medication Administration: the right medication, to the right patient, giving the right dose via the right route at the right time. The Five Rights have proven to be a good rule for avoiding medication errors, and what is good for medication administration is also good medicine for incident reporting. The following are the Five Rights of Incident Reporting:

- The Right Form
- The Right Facts
- The Right Focus
- The Right Forum
- The Right Follow-Up

The Five Rights of Incident Communication
The Right Form

Incidents must be documented using the appropriate form. First, let's begin with establishing appropriate terminology. The terms *incident report* and *QA Report* are both outdated and have a negative connotation. If seen by outsiders of EMS, anything titled with the word *incident* or *quality* may communicate the appearance of negligent action. Some states protect internal communication, but others may leave the incident report discoverable for legal purposes. Hence, EMS providers should avoid the terms and instead use the term *Internal Communication Report (ICR)* (which is used for the remainder of this discussion).

Choosing the right form to document an incident ensures appropriate action or resolution. A general rule is to record patient care alone on the Patient Care Report. If an equipment problem occurs during the course of providing patient care, document the care given to the patient in the PCR, but leave the details of the issue for the Internal Communication Report. If an error was made during the course of treating a patient, accurately document the *treatment* given to the patient in the PCR and then document the *event* in the ICR.

The Right Facts

The right facts must be presented in the ICR. Incident reporting is for the internal communication of facts related only to the issue at hand. Because the report must be objective, it is not the place to record gripes, complaints, or your own opinions and conclusions or to serve as a therapy session for a frustrated EMS professional. The importance of this cannot be overstated.

The Right Focus

The ICR must have the right focus. Incident reporting should be focused solely on communicating the problem or issue at hand. Emotions can motivate an EMS professional to document an incident. Ask yourself, what is the focus of this report? As you document, be careful not to get off subject; stay on target and stay focused.

The Right Forum

risk management
A function of leadership within an organization that seeks proactively to identify, manage, and reduce the risks associated with everyday EMS activities.

The ICR must have the right forum. Because incident reporting is very serious, it is important for you to direct the report to the most appropriate person(s). The purpose of incident reporting is for communication between clinical staff and administration, primarily for quality and **risk management** purposes. The right forum helps to ensure that the ICR arrives at the desk of the person who can provide the most appropriate corrective action in a timely manner. Table 11-1 provides a summary of forms and forums for incident reporting.

Table 11-1	Incident Report Forms and Forums
Type of Issue	**Form/Forum**
Patient care delivery	Patient Care Report (PCR)
Actual patient care	Patient Care Report (PCR)
Patient care problems	Internal Communication Report (ICR)
Partner problems not affecting patient care or delivery	See your supervisor
Dispatch issues not affecting or related to patient care or delivery	See your supervisor
Equipment issues	Internal Communication Report (ICR)

The Right Follow-Up

The ICR must have the right follow-up. Once you have documented and submitted the Internal Communication Report, it becomes someone else's responsibility to ensure proper follow-up.

General Guidelines for Internal Communication

General guidelines for the Internal Communication Report include:

- If the incident involved an equipment failure, document the name, model, and serial number of the equipment. Make sure to follow your organization's procedures for ensuring the device is not placed back in service. After doing so, document that this was done.
- The ICR is for administrative use. Keep the documentation in the PCR separate from any incident reporting. Do not reference the ICR in the PCR.
- The basic principles of documentation also apply to incident reporting. The report must be clear, complete, correct, consistent, and concise.
- Complete the report as soon as possible after the event and promptly submit the form to your direct supervisor.
- Include a descriptive narrative summarizing the incident and your immediate action in response to the issue.
- If patient care was involved, reference the PCR and provide a description of any applicable patient responses, including statements made by the patient.
- Make sure that all persons who witnessed the event are noted in the report.
- Remember: The purpose of the report is to communicate an event so that proper investigation and resolution can take place.

These principles will guide the EMS professional through the process of incident report documentation. Be sure always to follow the standard operating policies and procedures of your organization.

Summary: Return to Case Study

Instinctively, you turn the patient on his side and begin clearing the patient's airway utilizing BLS maneuvers. Your partner uses the elevator emergency phone to alert hospital security that you have a patient in cardiac arrest. While the elevator ride down to the first floor seems like an eternity, within seconds the door opens and the code team from the ED is racing toward you with a crash cart. Later, after you have completed the PCR, contact the on-duty supervisor and complete an Internal Communication Report.

The Internal Communication Report is a valuable and necessary communications tool for incident reporting and is the preferred method for the written communication of incidents unrelated to patient care. In documenting incidents, it is important for the EMS professional to document using the right form, recording the right facts with the right focus, and ensuring that the incident is being addressed to the right forum so that the right follow-up can take place.

CHAPTER REVIEW

Review Questions

Refer to Answers to Chapter Review Questions at the back of this book.

1. What type of events should be documented utilizing the Internal Communication Report?

2. List the Five Rights of Incident Reporting.

3. What role does incident reporting have in risk management for EMS providers?

4. List and describe the general guidelines for incident reporting.

Critical Thinking

Please refer to Answers to Critical Thinking Discussion Exercises at the back of this book.

1. A patient was dropped while being transferred from the hospital bed to the ambulance stretcher. Evaluate the following statement on the EMS provider's Incident Communication Report: "The patient fell while being lifted from the hospital bed onto the ambulance stretcher. The patient fell because the nurse failed to lock the wheels on the hospital bed."

Action Plan

1. Apply the Five Rights of Incident Reporting to your daily practice.

2. If you do not have safety, patient care, or process improvement committees in your organization, take the lead and suggest that they be established.

3. Whenever you document an event, always seek feedback as to the resolution of the problem.

Practice Exercises

1. Evaluate the Internal Communication Report regarding equipment malfunction in Figure 11-2.

   • What are the problems with the documentation in this ICR?

   • How could this report be rewritten using the principles presented in this chapter?

EMS Documentation Internal Communication Report	
Date: *June 15, 2007*	Report Number: *2005000015*

Report Type:

(Equipment)	Accident/Injury	Patient Management Error
Protocol Deviation	Other: _____	

Report Submitted By: *J. Smith*	Report Completed By: *J. Smith*
Time of Incident: *1515*	Time Report Completed: *1545*

Witnesses to Event:
1. *R. Jones*   Title: *Emergency Medical Responder*
2. *T. Deux*   Title: *AEMT*

Supervisor:
Name: *B. Hartley*   Time Notified: *1530*   Time Responded: *1530*

Equipment:
Equipment Type: *Portable Suction*
Manufacturer: *Acme*   Model Number: *B123*   Serial Number: *107*
Control Number: *12345*   Equipment Impounded: Y   N   Time: *1545*
Equipment Tagged: Y   N   Time: By: *B. Hartley*

Accident/Injury:
Accident Type:

Patient Management Error:
Type of Error:
Cause:

Protocol Deviation:
Protocol:
Reason:

Narrative:
Called to residence for patient found in cardiac arrest. On arrival, patient noted to be pulse-less and apneic. Prepped for intubation: high-flow O2 and oral airway. On first attempt at intubation, patient vomited. Attempted to suction airway but suction unit failed. Emergency Medical Responders failed to bring their suction unit in with them on this call. Documented event on PCR. Unable to resuscitate patient.

Signature:	Title	Date/Time
1. *John Smith*	*Paramedic*	*6/15/2007 1615*
2. *Ted Deux*	*AEMT*	*6/15/2007 1615*

Administration Use Only:
Received By: Date/Time:

FIGURE 11-2
Equipment Malfunction Report

2. Evaluate the Internal Communication Report regarding medication error in Figure 11-3.

- What are the problems with the documentation in this ICR?

- How could this report be rewritten using the principles presented in this chapter?

EMS Documentation Internal Communication Report	
Date: *11/05/2007*	Report Number: *2004000156*

Report Type:

Equipment	Accident/Injury	(Patient Management Error)
Protocol Deviation	Other: _____	

Report Submitted By: *A. Smith*	Report Completed By: *A. Smith*
Time of Incident: *0600*	Time Report Completed: *1800*

Witnesses to Event:
1. *A. Smith*   Title: *Paramedic*
2.   Title:

Supervisor:
Name: *J. Jordan*   Time Responded: *1900*
Time Notified: *1830*

Equipment:
Equipment Type:
Manufacturer: Model Number: Serial Number:
Control Number: Equipment Impounded: Y   N   Time:
Equipment Tagged: Y   N   Time: By:

Accident/Injury:
Accident Type:

Patient Management Error:
Type of Error: *Medication*
Cause: *Label*

Protocol Deviation:
Protocol:
Reason:

Narrative:
Patient was to have been administered 2 mg of MS, but misread label and gave patient 5 mg of MS. Supervisor notified.

Signature:	**Title**	**Date/Time**
1. *A. Smith*	*Paramedic*	*11/05/2007*
2.		

Administration Use Only:
Received By: Date/Time:

FIGURE 11-3
Medication Error Report

3. Evaluate the Internal Communication Report regarding patient injury in Figure 11-4.

- What are the problems with the documentation in this ICR?

- How could this report be rewritten using the principles presented in this chapter?

EMS Documentation Internal Communication Report	
Date: *01/27/2006*	Report Number: *20060011*

Report Type:

Equipment	Accident/Injury	Patient Management Error
Protocol Deviation	Other: *Patient Injury*	

Report Submitted By: *R. Wood*	Report Completed By: *R. Wood*
Time of Incident: *0810*	Time Report Completed: *0810*

Witnesses to Event:
1. *C.R.* Title: *RN*
2. *T.B.* Title: *RN*

Supervisor:
Name: *S. Nychols*
Time Notified: *1000*   Time Responded: *1100*

Equipment:
Equipment Type:
Manufacturer: Model Number: Serial Number:
Control Number: Equipment Impounded: Y N Time:
Equipment Tagged: Y N Time: By:

Accident/Injury:
Accident Type: *Patient injury*

Patient Management Error:
Type of Error:
Cause:

Protocol Deviation:
Protocol:
Reason:

Narrative:
Patient fell while being transferred from stretcher to hospital bed. The nurse failed to lock the wheels on the hospital bed properly. Patient fell to ground, striking head. Supervisor notified, but significant delay in response. Patient assessed. Alert and oriented X4 w/no neuro deficits. Approx. 2 inch laceration to forehead. Placed on LSB and transported to ED for evaluation.

Signature:	Title	Date/Time
1. *R. Wood*	*EMT*	*01/28/2006 0800*
2.		

Administration Use Only:
Received By: Date/Time:

FIGURE 11-4
Patient Injury Report

CHAPTER 12

Verbal Reports

Key Ideas

Upon completion of this chapter, you should know that:

- Each part of the health care continuum is dependent upon reporting and documentation from the preceding care providers and is essential for continuity of care.

- Legal considerations in verbal reporting center upon protecting patient confidentiality and inadequate transfer of care.

- The information provided in the Pre-Hospital Report will enable the physician to prepare for lifesaving invasive interventions.

- The EMS DATA report format provides a basic structure for organizing a Pre-Hospital (radio) Report.

- Nursing professionals require information relating to choreography, how the patient fits and moves into the hospital system, matching patient needs with appropriate resources.

- The DATA RN format provides a basic structure for organizing a Transfer of Care Report.

FIGURE 12-1
(Courtesy EMSA, Tulsa, OK)

Y ou are inbound emergency status to the nearest trauma center with a 57-year-old female patient from a motor vehicle collision. She was the lone occupant/driver of a vehicle that was struck broadside in a busy intersection at a high rate of speed, producing approximately 12 inches of intrusion into the patient compartment. She was not wearing safety restraints and air bags were not deployed. Currently the patient is alert and oriented to person, place, time, and event, and is complaining of severe abdominal pain, rated 6:10. She denies any loss of consciousness and is able to recall the incident in detail. Vital signs are blood pressure—90/40, heart rate—120, and respiratory rate—16.

Trauma assessment is negative except for ecchymosis to the left side of the abdomen and notable tenderness and rigidity to the left upper abdominal quadrant. Past medical history includes an episode of diverticulitis in 1990 and hypertension for which she takes torsemide and Toprol. Other medications include Premarin, numerous herbal supplements, two aspirin per day ("because of those commercials on TV"), and Prozac.

EMS treatment includes full spinal immobilization, oxygen 15 L per non-rebreather, and cardiac monitoring, which shows a sinus tachycardia. You have an IV of normal saline established that is infusing at 100 ml/hr. You anticipate arriving at the hospital in 10 minutes. Five minutes out from the hospital you give the following radio report:

> EMS 3 to MRMC, we are inbound with a 57-year-old female patient from an MVC. Pt is alert and oriented. Vital signs are BP 90/40, heart rate 120, and respiratory rate 16. She is on O_2 15 liters per non-rebreather. Cardiac monitor shows a sinus tachycardia. The patient had diverticulitis in 1990 and takes aspirin and Toprol. We have established an IV of normal saline that is running at 100 ml/hr. Our ETA to your facility is 5 minutes. Do you have any questions or orders?

Once in Trauma Room One at the Emergency Department, you begin giving a report to the nurse as you move the patient from the stretcher to the examination table:

> This is the 57-year-old female who was in the MVC. She is complaining of abdominal pain to the left side. She is alert and oriented X4.

Later, after you finish the PCR, you remember that you forgot to tell the nurse that the patient takes Prozac for depression. As you walk past the nurse's desk, you state, "By the way, Trauma One takes Prozac for depression."

Questions

Please refer to Answers to Case Study Questions at the back of this book.
1. Did the radio report provide the appropriate information to the receiving hospital? Why or why not?

2. Was the information provided in the verbal report to the RN at the hospital appropriate?

3. Evaluate the manner in which the nurse was advised of the "additional medication." Was this appropriate?

Introduction

continuity of care
Seamless transitions in the delivery of health care from one provider, or department, to another.

Each part of the health care continuum is dependent on reporting and documentation from the preceding care providers and is essential for **continuity of care**. The Emergency Department is dependent upon EMS for proper reporting and documentation. The ICU is dependent upon the ED, surgery is dependent on the ICU, and so on. If any one of the players in continuity of care fails to document its care and treatment, the next decision makers will lack critical information and deadly errors can take place. Verbal reporting and the PCR are essential to continuity of care.

Verbal reporting can be frustrating. On one hand, there is the desire to give hospital staff the information needed to initiate hospital treatment, but on the other hand, one can wonder whether EMS reports are given proper attention. Radio reports, contacting medical control for physician orders, and giving the verbal report to the nurse who is receiving the patient can be intimidating for those new to EMS. Knowing what information is required from EMS is essential for effectiveness in verbal reporting. This chapter will examine the important role of verbal reporting in the transfer of care and its link to PCR documentation.

ON TARGET Each part of the health care continuum is dependent on reporting and documentation from the preceding care providers and is essential for continuity of care.

KEY TERMS

Note: Page numbers indicate where the following key terms and definitions first appear.

continuity of care (p. 175)   acuity (p. 178)

Legal Issues and Verbal Reporting

Two primary legal considerations are associated with verbal reporting. First, there are confidentiality concerns associated with any form of patient communication. The Health Insurance Portability and Accountability Act of 1996 (HIPAA) provides legal protection for the security and privacy of patient health information. Therefore, HIPAA impacts the manner in which confidential patient information is communicated in verbal reporting. In any dialog about a patient's health information, risk of a HIPAA violation is present. Consider the following guidelines for protecting confidentiality in verbal reporting:

- Anytime a report can be heard by someone who will not be taking care of the patient, there is a potential breach of patient confidentiality. Make sure that you are giving your report to the actual person assuming responsibility for the patient.
- Avoid radio reports, if at all possible. Cell phones offer better protection of patient's protected health information.
- Privacy is the rule. Therefore, as much as possible, give and take Transfer of Care Reports in private away from anyone not needing the information. Give a Transfer of Care Report only to the nurse or physician who is assuming responsibility for the patient.
- Remember, HIPAA must never compromise patient care.

Second, inadequate care transfer presents a legal risk to the EMS professional. The EMS provider must be accurate in verbal reporting so that appropriate information is provided for the legal exchange of transfer of care actually to occur. If you blab a quick five-second report as you are running out the door for the next call (or the next coffee), you have failed to transfer care appropriately. Failure to transfer responsibility for the patient's care appropriately to another qualified health care professional could be considered abandonment, and the EMS provider could be found negligent if there were a poor outcome.

Verbal Reporting and the PCR

There are three stages of the EMS transfer of care:

1. **The Pre-Hospital Report:** The Pre-Hospital Report, or radio/telephone report, alerts the receiving facility and introduces the patient to the next care providers. The Pre-Hospital Report provides only the basic information required to prepare for the patient's arrival.
2. **The RN Transfer of Care Report:** The RN Transfer of Care Report builds upon the Pre-Hospital Report and provides all necessary information for the receiving health care professional to assume responsibility and continue care. The RN Transfer of Care Report is *not* the same as the Pre-Hospital (radio) Report even though some EMS personnel mistakenly give the same report to the nurse as was given over the radio or phone.
3. **The Patient Care Report (PCR):** The PCR provides comprehensive written summarization of the entire EMS encounter as well as documented affirmation that transfer of care occurred.

Transfer of care is the exchange of responsibility from the EMS professional to (usually) the nursing professional. As you give the essential information regarding the patient, the responsibility for the patient passes to another. The receiving hospital:

- Must know the patient's condition for which he or she is being treated.
- Must know patient demographics.
- Must know the patient's pertinent past medical and social histories.
- Must know the treatment that has been given prior to transfer of care.
- Must be able to ascertain from the Transfer of Care Report the priorities for the patient.

The Pre-Hospital Report, the RN Report, and the PCR each have a distinct purpose but fit together like gears in the transmission driving the vehicle of continuity of care. All three reports must be in agreement in providing the same impression of the patient's condition. Together, the Pre-Hospital Report, the RN Transfer of Care Report, and the Patient Care Report ensure continuity of care. Table 12-1 summarizes the purposes and characteristics of these reports.

Table 12-1 Transfer of Care Reporting

Report	Purpose	Characteristics
Pre-Hospital (Radio/Phone) Report	Introduces the patient to the receiving facility for the purpose of preparation	• Brief • Follows EMS DATA format • Focused on information needed to prepare for care
RN Transfer of Care Report	Introduces the patient to the next health care professional for the purpose of assuming legal responsibility and initiating care	• Detailed, but focused • Focused on information needed to continue care • Communicates priorities • Follows DATA RN format
Patient Care Report (PCR)	Documents all aspects of the EMS event	• Comprehensive written report • Validates that proper transfer of care has occurred

The Pre-Hospital (Radio) Report

The radio report serves an important function in EMS, tying pre-hospital to hospital care, and represents the beginning of the transition out of EMS to in-hospital care. The radio report communicates the information to the receiving hospital that will best assist its staff in preparing for the patient. It enables the hospital to begin preparations for continuing care and the hospital's decision-making process regarding establishing a plan of care prior to the arrival of the EMS patient at its facility.

A long-standing complaint in EMS has been that nursing staff at receiving hospitals do not pay proper attention to EMS radio reports. Have you ever noticed they generally are ready for the "big calls," the traumas and cardiac arrests, but not for the lower-priority calls? A recent study, published in *Prehospital Emergency Care*, evaluated whether the information provided by EMS providers in the radio reports for emergency and nonemergency patients resulted in preparatory action. The study concluded that radio reports rarely give information to the Emergency Department that is acted upon:

Radio reports for low-priority patients may not be an efficient or productive use of providers' or nurses' time.[1]

This study details how EMS radio reports are perceived by ED staff and suggests that receiving facilities are triaging EMS radio reports. Perhaps Emergency Department staff are not being provided the appropriate information from EMS radio reports. They will listen if the information is valuable to them.

Reporting to the Physician

Generally, when you are giving a report to a physician, it is for medical control, typically for the purpose of obtaining an order for a medication or intervention. There is certainly a right way and a wrong way to discuss patient care with a physician. The physician needs only the information pertinent to enable him or her to make a decision and either proceed with or deny the order. The physician needs:

- Brief demographics.
- Presenting illness/injury that makes the patient stable versus unstable.
- Pertinent past medical history that relates to the condition and order being requested (not the full medical history).
- Vital signs and specifics as to instability or why the order is requested.
- Specific order being requested.

The information provided in your report will enable the physician to prepare for life-saving invasive interventions. The physician wants to know whether his or her attention will be needed immediately upon the patient's arrival. Is the patient unstable? Does the patient have a patent airway, and is the patient's cardiac/respiratory status stable? The physician also needs to know patient information that will enable him or her to:

- Facilitate physician specialists such as anesthesiology, surgery, or specialized interventional services.
- Prepare for invasive procedures such as RSI or surgical airway.
- Prepare for emergent transport (such as air medical) if services are not available within the receiving medical facility.

[1] M. S. Penner, D. C. Cone, and D. MacMillan, "A Time-Motion Study of Ambulance-to-Emergency Department Radio Communications," *Prehospital Emergency Care*, 7, 2(2003), 204–208.

Most physicians respect EMS. By giving physicians focused information in a professional, confident manner, they will be attentive and you will be more effective in obtaining the orders needed to care for your patients optimally. Be sure to know the level of licensure of the person you are giving a radio report to. "Do you have any orders?" should never be asked in a radio report unless you know you are speaking with a medical control physician.

Pre-Hospital Report Formats

Every EMS agency should have a uniform radio report format. Many EMS systems mandate a specific format for radio reports, whereas others designate a certain format for certain medical or trauma conditions. For instance, some EMS systems mandate a particular, more concise format for activating certain trauma or cardiac criteria. For example, "Unit 11—We are inbound with a 'STEMI,' ETA 10 minutes," or "Unit 11—we are inbound with a 'trauma red,' ETA 10 minutes."

If your EMS system does not mandate a designated format, consider using the EMS DATA format as a base structure for organizing radio and phone reports:

EMS = Service/Unit, Hospital and Incident Number, ETA
D = Patient Demographics
A = Chief Complaint and Assessment
T = Treatment
A = Acknowledgment

Figure 12-2 provides an example of a Pre-Hospital Report using the EMS DATA format.

The RN Transfer of Care Report

The RN Transfer of Care Report is absolutely essential to the continuity of care. Its purpose is to continue the process started in the Pre-Hospital Report, which introduced the patient to the receiving facility. The RN Transfer of Care Report serves two vital functions: it provides the detail the nurse will need to continue care, and it introduces the patient to the next health professional who will be continuing care.

Reporting to the RN

acuity
The complexity of a patient's condition.

The informational needs of the physician and the nurse are different. Understanding how to give a report to the RN is found in understanding the world of nursing. EMS professionals generally care for one patient at a time. The RN may have responsibility for two, three, or four high **acuity** patients at a time, in addition to carrying out physicians' orders and providing oversight to technical staff. If you fail to give the RN the information needed in the first 10 seconds of your report, you have probably lost him or her. RNs simply don't have the time for a play-by-play account of the EMS call.

EMS: County EMS Unit 3 to University reference 20074123. Inbound ETA 5 minutes emergency status. **D**: Patient is a 25-year-old alert/oriented female **A**: Complaining of sharp, severe lower abdominal pain progressive × 12 hours, vaginal bleeding, and one syncopal episode. She is pale/diaphoretic with BP = 70/40, HR = 120, RR = 16. **T**: Treatment is supportive. IV access has been obtained with normal saline infusing at 200 ml/hr. **A**: University, do you acknowledge report?

FIGURE 12-2
EMS DATA Pre-Hospital Report Example

The registered nurse is motivated by patient placement and patient flow through the hospital care system. The nurse requires information that will enable him or her to:

- Continue care and treatment immediately.
- Assign a room/bed in the appropriate ED unit. Does the patient need a high-priority ED room or a lower-priority ED room?
- Arrange for placement as an inpatient if a direct admit.
- Assign the appropriate acuity level for the patient.
- Anticipate the interventional needs of the patient.

The nurse decides how best to fit the EMS patient into the system. Therefore, the following information from the Pre-Hospital Report is needed:

- Trauma/medical—stable/unstable.
- Demographics for room assignment.
- Patient's chief complaint/EMS Diagnosis.
- Patient's physician.
- EMS treatment.
- What resources will be needed (respiratory therapy, security, and so on)?

Physicians require information that will enable them to prioritize and prepare for interventions. In addition to continuing care, nursing professionals require information relating to choreography, how the patient fits and moves into the hospital system, matching patient needs with appropriate resources.

Guidelines for Transfer of Care Reporting

A proper transfer of care is dependent upon communicating the appropriate information in the proper manner. The EMS professional should use the following principles in transfer of care reporting.

1. **Know to whom you are talking:** Who is this person from whom you are taking a report? Is this the patient's primary nurse, or is the patient's nurse at lunch (warning sign!)?
2. **Do not rely on your memory:** Use the PCR as your guide in transfer of care reporting.
3. **Focus on priorities:** Clue the RN in as to the patient's priorities, instead of just giving a play by play of the EMS event.
4. **Introduce the patient:** Use the Transfer of Care Report to introduce your patient to the next caregiver. Give the nurse the report in the presence of the patient. This will assist the nurse in being attentive to the report and will bring the patient into the exchange.
5. **Summarize:** End the report with a summary of the EMS Diagnosis, the care that you gave, and the priorities for the patient.
6. **Demand attention:** Look for cues that the nurse is plugged into the exchange. Are you talking to yourself, or is someone else involved in the conversation? Is there eye contact? Does the head nod? Is the nurse asking questions? If you find the nurse disengaged from the process, probe further.
7. **Demand feedback:** Encourage the receiving nurse to acknowledge that he or she understands what you have communicated.
8. **Avoid medical clichés:** Don't use clichés such as "She's a stable cardiac." Instead, be specific. What makes the patient stable or unstable?

Transfer of Care	
Receiving Facility: *Regional Medical Center*	Destination Type: *Hospital*
Condition of Patient at Destination: *Improved/Stable*	
Pre-Hospital Report to Destination Facility Time: *1600* Via: *Cell Phone*	Person: *Tim Jones, RN*
Medical Records (Type): *Medication List*	Received By: *Tim Jones, RN*
Patient Belongings (Type): *None*	Received: *Tim Jones, RN*
Transfer of Care: I have received an appropriate Transfer of Care Report: Time: *1615* Signature: *Tim Jones, RN*	Title: *RN* Medical Records Received: *TJ*

FIGURE 12-3
EMS Documentation PCR—Transfer of Care

9. **Ask questions:** If you are receiving a report at a sending facility, don't be satisfied until you "know" the patient.
10. **Document the transfer of care on the PCR:** Documentation of transfer of care completes the PCR. Figure 12-3 provides an example of validation of transfer of care using the EMS Documentation PCR.

RN Transfer of Care Report Formats

Transfer of Care Reports should also utilize a standardized format. Transfer of care formats should be based upon the PCR used by your EMS system. Having a report format that is used universally by your EMS organization will:

- Provide structure for delivering reports.
- Train the nursing staff at facilities in your area. By hearing reports in the same format, they will cue into important information.
- Provide for consistency.

If your EMS system does not mandate a designated format, consider using the DATA RN format for Transfer of Care Reports:

D = Patient Demographics
- Name and age
- Social history
- Resuscitation status
- Past medical history
- History of present event

A = Chief Complaint and Assessment Findings

T = EMS Treatment

A = Action Items/Priorities

R = Receiving MD/Sending MD

N = Nursing Priorities

Figure 12-4 provides an example of transfer of care utilizing the DATA RN format.

D—This is Mr. Smith. He is 75, lives at Forest Assisted Living Center, and is retired from government. His son lives locally and has power of attorney. He has an advance directive that specifies DNR. Past medical history is significant for COPD/Emphysema, and insulin-dependent diabetes.

A—Mr. Smith describes a two-hour exacerbation of respiratory distress, unrelieved by his albuterol inhaler, which usually relieves his dyspnea. He is alert and oriented to person, place, time, and event. Initial and last vital signs–BP = 142/88, HR = 110, RR = 32, O_2 saturation 82% on arrival. Last–BP = 126/84, HR = 100, RR = 16, O_2 saturation 92%, Temp = 102.1. Denies chest or other pain, Breath Sounds–diminished left side, with crackles in right base. Productive cough with yellow sputum.

T—Our treatment–Oxygen 4 L via nasal cannula, cardiac monitoring–sinus rhythm, two nebulized albuterol treatments with decrease in dyspnea and increase in O_2 saturation. No side effects were noted from the albuterol. IV access obtained in right forearm with a 20 gauge catheter. The normal saline is infusing at TKO, and the fluid total is 50 ml.

A—Mr. Smith ate lunch but did not receive his 40 units of regular insulin at 1300.

R—Sending and receiving physician is Steve Johnson.

N—Mr. Smith's nebulized treatments seem to last about 12 minutes, and it has been 10 minutes since he finished his last breathing treatment. He does seem to desaturate quickly. His family has not been notified. Please see PCR for current medications and complete information relating to his past and present medical histories, our assessment and treatment. Do you have questions?

FIGURE 12-4
DATA RN Transfer of Care Example

Summary: Return to Case Study

As you head back to the ambulance, the patient's nurse runs out to you and states: "I know you're probably in a hurry to get back to your base, as I was a medic once, but I really need more information about this patient. Can you come back in and give me some additional information? I would greatly appreciate it."

Proficiency in verbal reporting is an essential process in the continuity of care. The three stages of patient transfer—the Pre-Hospital Report, the RN Transfer of Care Report, and the Patient Care Report (PCR)—move the patient from EMS care into the health care system. The EMS DATA and the DATA RN formats provide the EMS professional a structure in order to communicate effectively the information that is vital to the transfer of care.

CHAPTER REVIEW

Review Questions

Please refer to Answers to Chapter Review Questions at the back of this book.

1. List and describe the two legal concerns associated with verbal reporting.

2. List and describe the three stages of EMS transfer of care.

3. Describe the purpose of the Pre-Hospital (radio) Report.

4. Describe the purpose of the RN Transfer of Care Report.

5. List and describe the purposes and characteristics of the Pre-Hospital Report, the Transfer of Care Report, and the PCR.

6. Describe the differences in reporting to the physician and the registered nurse.

7. List the elements of the EMS DATA Pre-Hospital Report format.

8. List the elements of the DATA RN format for Transfer of Care Reports.

Critical Thinking

Please refer to Answers to Critical Thinking Discussion Exercises at the back of this book.

1. What are the advantages of mandated report formats for verbal reporting?

2. Why is it helpful for the EMS professional to identify nursing priorities in the Transfer of Care Report?

Action Plan

1. If your EMS system records radio reports, ask to listen to one or two of your reports. Evaluate them using the principles presented in this chapter.

2. If your EMS system does not mandate specific formats for verbal reporting, consider using the EMS DATA and DATA RN formats.

Practice Exercises—Role Play

Using the EMS DATA and DATA RN formats, organize the information in the scenario for transfer of care reporting. Then, using role play, practice giving the Pre-Hospital and RN Transfer of Care Reports, applying the principles presented in this chapter. If possible, record the reports for critique and evaluation.

Scenario Information

Patient is a 60-year-old male with chest pain. Past medical history includes hypertension and acid reflux disease. Patient states he had gallbladder surgery 5 years ago and is allergic to penicillin. He is alert and oriented to person, place, time, and event. Pain began at rest, is substernal, but has no radiation. Patient denies any stressors or events that provoked his pain. He is also allergic to Novocain. He lives with his wife, who is en route to the hospital. No advance directive. Medications include Prilosec and Toprol. Skin is warm and dry. Breath sounds are clear, and abdomen is soft. No peripheral edema. He states a previous episode of chest pain like this happened a month ago. No physician in the area as he just moved to the area after a divorce. 12-Lead EKG shows a sinus rhythm with ST depression in the anterior leads. Treatment: O_2 @ 12 L per non-rebreather, IV of normal saline in left forearm with 20 gauge catheter, at TKO. Vital Signs—BP = 140/90, HR = 60, RR = 12, Oxygen saturation—99%. Sublingual NTG—two doses en route, the last dose 5 minutes ago. Initial pain 8:10. Now 5:10. Order for morphine declined by medical control.

CHAPTER 13

Complex Patient Encounters

Key Ideas

Upon completion of this chapter, you should know that:

- EMS encounters involving crime scenes, behavioral crises, and issues of patient rights are difficult for the EMS professionals and place greater demands on documentation ability.

- Documentation of EMS encounters involving a crime scene victim will usually become part of the criminal investigation and judicial process.

- EMS management of the patient from a crime scene includes maintaining and preserving the integrity of the crime scene in practice and in the PCR.

- Upon entry into the crime scene, the EMS professional's most valuable assessment tools are not in the first response bag, but the senses of sight, sound, and smell.

- The patient's social history provides valuable insight into his or her environment, living conditions, and social interaction.

- One of the most critical aspects of the management of the sexual assault victim is the preservation of evidence, which must be reflected in PCR documentation.

- Documentation of the management of the victim of child and elder abuse must capture essential elements from the patient's environment noted by the EMS professional.

- Today's EMS professional must assume the role of **patient advocate** in assuring that advance health care decisions are appropriately honored.

patient advocate
An advocate is someone who acts on behalf of another.

CASE Study _____

You are dispatched to the scene of a potential assault involving a child. Upon arrival you find an 8-year-old boy being cared for by Emergency Medical Responders, who report that neighbors summoned police after hearing a child screaming. Police officers found the child submerged facedown in a bathtub and immediately placed the child's caretaker into custody after calling for EMS.

The patient, covered in a blanket, remains wet. He is conscious but oriented to person only. The skin is pale with mild cyanosis around the mouth and nail beds are also cyanotic. Vitals signs are blood pressure—80/65, heart rate—120, and respiratory rate—32. Oxygen saturation is 85 percent. Your partner applies high-flow oxygen via non-rebreather as you

FIGURE 13-1
(Courtesy of EMSA, Tulsa, OK)

proceed with the remainder of your assessment and examination. Most notable is the impression of a handprint on the top of the child's head.

You and your partner provide emergency treatment and transport to the closest trauma center. Later, in the EMS report room, you are approached by a police detective who states the suspect in custody is being charged with attempted murder and a criminal investigation is underway. The call was challenging in itself but now the thought comes to mind, how will I document this?

Questions

Please refer to Answers to Case Study Questions at the back of this book.

1. Are complex patient encounters any different from other "everyday" EMS encounters? Why or why not?

2. If you were advised by a law enforcement officer that a patient encounter were going to involve a criminal investigation, would it change your approach to PCR documentation? Why or why not?

Introduction

The case study illustrates one of the "tough calls" to which EMS professionals are called. Patient encounters involving crime scenes, behavioral crises, and issues of patient rights are difficult for the EMS professional. PCR documentation of these complex patient encounters generally becomes part of the judicial process. Therefore, greater demands are placed on

There must never be a double standard in documentation, one for the "easy" encounters and another for the "tough" encounters.

the EMS professional's documentation ability in order to capture information critical to the protection of patient rights.

The principles of documentation presented in this text apply to all patient encounters, regardless of complexity. There must never be a double standard in documentation, one for the "easy" encounters and another for the "tough" encounters. This final chapter will focus on capturing the additional critical information that is essential to both the judicial process and the rights of the EMS patient.

KEY TERMS

Note: Page numbers indicate where the following key terms and definitions first appear.

patient advocate (p. 185)

advance directive (p. 192)

living will (p. 192)

health care proxy (p. 192)

durable power of attorney for health care decisions (p. 192)

EMS Documentation and the Crime Scene

EMS patient encounters involving crime scenes are almost always challenging situations, both personally and professionally. Because PCR documentation will become part of the criminal investigation and judicial process, documentation of patient encounters involving a crime scene victim requires additional considerations. Applying the following principles to PCR documentation enables the EMS professional to assist, through the PCR, in the judicial process.

General Principles of Crime Scene Documentation

The patient from the crime scene requires a different approach not only in scene management but also in documentation. In certain instances the EMS professional is the first public safety representative to arrive on the scene. Law enforcement will depend on the EMS professional to provide valuable crime scene information to enable the judicial process to work for the crime victim.

In providing EMS care and removing the patient for transport, the crime scene is disturbed. The PCR may be called upon to assist in re-creating the crime scene at the point of EMS entry. Without compromising care, the EMS professional must manage the patient as well as attempt to manage the integrity of the crime scene. Remember your training. Preserve the scene but also preserve it through the PCR. Essential elements of crime scene documentation include:

1. **The Crime Scene:** Upon entry into the crime scene, the EMS professional's most valuable assessment tools are not in the first response bag, but the senses of sight, sound, and smell.

   Sight: What did you see?
   - If immediately apparent, describe the mechanism of injury. Figure 13-2 provides an example of descriptive documentation for mechanism of injury.

Narrative Snapshot: Mechanism of Injury

Noted a long barrel handgun lying approximately 5 feet distal to the patient's lower extremities. Handgun had a white "mother of pearl" handgrip. A trigger lock was noted on the bedside table located on the left side of the bed as you enter the room.

FIGURE 13-2
Mechanism of Injury

FIGURE 13-3
Patient Presentation

- Document the exact location and description of how the patient was found upon arrival. Figure 13-3 provides an example of descriptive documentation for patient presentation.
- If EMS is first on scene, who was present at the scene? Document everyone you note to be on scene, especially those within close proximity to the patient.
- What objects were present at the scene? If objects pertinent to the injury or illness are noted, document a description of the object(s) and the exact location found. If during the course of EMS treatment any objects (furniture, etc.) were moved, describe the item, the original location, and where it was moved.
- If any objects were removed from the patient (such as jewelry or other personal items), document each item, its original location, and where it was placed. If the item was given to someone, document who received the item.
- If any patient clothing was removed, document the article of clothing, how it was removed, and where it was placed. If the item of clothing required cutting in order to be removed, document how the clothing was cut in order to preserve tears or cuts caused by entrance or exit wounds.
- Document any medical supplies, packaging materials, or other EMS article left on scene.

Sound: What did you hear?

- Statements made by the crime scene patient and bystanders are crucial to PCR documentation of the crime scene. It is not uncommon for the patient or pertinent bystanders to provide EMS with relevant information to a criminal investigation. Therefore, document, as much as possible, exact quotations from the patient or bystanders relevant to the incident.
- If EMS is first on scene, did the patient (or bystanders) report unusual sounds? If this information is ascertained in the course of the assessment, it is crucial to document it in the PCR. Figure 13-4 provides an example of descriptive documentation in bystander statements.

Smell: Were there unusual odors?

- Document and describe any unusual scents or odors present. Odors indicating the presence of alcohol or drug manufacturing are essential elements of crime scene documentation. Figure 13-5 provides a descriptive example.

FIGURE 13-4
Sounds from the Crime Scene

FIGURE 13-5
Odors from the Crime Scene

Narrative Snapshot: Social History
The patient is a 35-year-old male who lives alone after a recent divorce. Building manager reports seeing no activity at the residence over the last five days, and his dog has been barking "day and night."

FIGURE 13-6
Social History

2. **The Social History:** The patient's social history provides valuable insight into his or her environment, living conditions, and social interaction. Although obtaining a patient's social history is not always appropriate or possible in the EMS environment, it can be useful information not only in the care continuum but also in the crime scene investigation. Figure 13-6 provides a descriptive example of social history documentation.

3. **The Assessment and Examination:** A thorough assessment and physical examination must be documented for all patients. The patient from the crime scene requires a higher level of descriptive documentation with respect to the patient's neurological status and traumatic injuries.

   - **Description of neurological status:** The patient's mental status upon EMS arrival is critical information that must be meticulously documented. If the patient was neurologically intact upon EMS arrival and later deteriorates, statements made by the patient will be crucial in law enforcement's investigation.
   - **Description of traumatic injury:** Because the patient's injuries are actual evidence, it is vitally important for the EMS professional to document a detailed description of wounds and injuries. Figure 13-7 provides an example.

The Sexual Assault Victim

The EMS professional's approach to the victim of sexual assault must be grounded in medical excellence, compassion, and insight to the unique needs of this patient group. One of the most critical aspects of the management of the sexual assault victim is the preservation of evidence, which must be reflected in PCR documentation. Essential documentation elements include:

- The victim is part of the crime scene. Document the appearance of the patient and the surroundings prior to assessment or interventions.
- Document any clothing that was removed, why it was removed, and the disposition of the item. Figure 13-8 provides an example.
- Document who was present with you during the course of evaluation, treatment, and transport.

Narrative Snapshot: Crime Scene Trauma
Exam revealed one wound to left upper chest: approximate 2 inch straight puncture wound, appears full thickness with no ecchymosis.

FIGURE 13-7
Crime Scene Trauma

Narrative Snapshot: Chain of Custody
No clothing was removed, except the patient's brown parka-type coat in order to assess pain and bruising to left shoulder. Placed in yellow bag and given to Officer Smith.

FIGURE 13-8
Chain of Custody

- Items that come into contact with the patient become evidence. This is particularly important with respect to sheets, blankets, and any bandaging materials used. Document the specific item(s), how they were used, and who received them during transfer of care. This protects the chain of custody with respect to criminal evidence.

The Child Abuse Victim

Child abuse can take a number of forms: physical, mental, and emotional abuse or neglect. The EMS professional provides medical care, provides emotional support to the child, and documents a comprehensive legal record of the incident. Essential documentation elements include:

- Document who was caring for the child at the time EMS was called.
- Document who was present when the child was interviewed and examined.
- Document the onset time of the alleged illness/injury versus the time EMS was called.
- Document the child's general appearance and state of health.
- Document any sequential events leading up to the injury and the calling for EMS.
- Document exact conversations (with quotations) you had with the parent or caregiver.
- Describe the interaction between the child and parent or caregiver. Figure 13-9 provides an example of documenting the interaction between the victim of possible child abuse and the care provider.
- Document the interview with the child, the questions asked, and how the child responded.
- Describe the child's injury, including location and characteristics (color, shape, and symmetry) of any markings on the child suggestive of injury.
- Document presence and characteristics of pain and the tool used for pain assessment.

The Elder Abuse Victim

Elder abuse is underrecognized as a form of abuse. EMS professionals must have an awareness of the characteristics of the elder abuse victim. Essential documentation elements include:

- Take careful note of the patient's surroundings. Are the utilities on? Does the residence seem too hot or too cold for the season? Does the patient have prescribed medications and adequate food? Document the findings.
- Document who was with the patient upon EMS arrival.
- Document the general appearance of the patient. Does the patient seem well nourished and cared for?
- Document any odors, such as the smell of urine and/or feces.
- Document the presence of any trauma or injuries.
- Document the interaction between the patient and family members or caregivers. How does the family or caregiver react to your questioning? Figure 13-10 provides an example.

The EMS professional must be objective in PCR documentation involving the crime scene patient. Documentation must be free of subjective opinions and conclusions. When emotions are ignited, it becomes easy to indict and convict in documentation. By meticulously and descriptively documenting the essential elements, PCR documentation will assist the legal process in bringing justice to the crime victim.

Narrative Snapshot: Suspected Child Abuse

No eye or physical contact noted between mother and child. Child responsive to grandparents on their arrival.

FIGURE 13-9
Suspected Child Abuse

FIGURE 13-10
Elder Abuse Victim

EMS Documentation and the Behavioral Crisis

EMS providers are often called to care for patients experiencing a behavioral crisis. These patients are typically on transport holds and will usually have additional orders for restraint devices during transport. Because transport holds and orders for physical restraints have significant legal ramifications, they require meticulous documentation. Essential documentation elements include:

- The order for transport hold and the physician's order for restraints must accompany PCR documentation, and the PCR should reference these attached documents.
- Document the specific manner in which the patient was restrained.
- Document neurovascular assessment per local practice guidelines.
- Document the manner in which the patient was educated regarding the transport hold and restraint procedures.
- Document the name of the physician ordering the hold or restraints and the time of the last physician and nursing assessments prior to departure.

The Suicidal Patient

EMS professionals are often called to render care to patients who either have expressed suicidal ideation or have made a successful or unsuccessful suicide attempt. Essential documentation elements include:

- Carefully document the patient's mood and affect. Describe the patient's surroundings and appearance in detail.
- Document the presence of any friends, family members, or bystanders at the scene.
- Document the presence of any suicide note and document the disposition of the note.
- If a suicide attempt was made, document the mechanism used. If the patient or the mechanism was moved, document the original location and where the object was relocated. In cases when the attempt has been successful, if pertinent objects relating to the attempt (medication bottles, weapons or rope, etc.) were not disturbed, make it clear in documentation that scene integrity was maintained. Figure 13-11 provides an example.

FIGURE 13-11
Suicide Scene

Complex encounters can easily draw the EMS professional into emotional involvement. Within proper boundaries, emotional involvement is beneficial to patient care because the EMS professional will communicate care for the patient's well-being. Compassion is a powerful force. Emotional involvement, however, must not be reflected in PCR documentation.

EMS Documentation and the Rights of Patients

Our society has evolved with respect to health care rights. Patients no longer are passive individuals waiting for others to make medical treatment decisions for them. They now are actively involved in their own health care choices and decision making. The EMS professional must assume the role of patient advocate by ensuring that advance health care decisions are appropriately honored. Of particular concern are issues associated with advance directives.

The PCR and the Advance Directive

You arrive on the scene of a cardiac arrest. A family member meets you at the door and states, "You won't need any equipment. She doesn't want anything done but the police department said that EMS has to come out anyway. Please, just take a look at her and call the funeral home." What will you do?

Competent patients have the right to be involved in their own health care decisions. The Patient Self-Determination Act of 1991 specifically addressed the issue of the **advance directive**. All hospitals receiving Medicare and Medicaid reimbursement are required to inform patients of their rights concerning advance directives. The advance directive is a document by which a patient directs his or her medical treatment in advance. There are three types of advance directives.

> **advance directive**
> An advance directive is a written statement directing a person's health care treatment in advance.

> **living will**
> A type of advance directive that stipulates the types of treatment that the patient desires to refuse (or accept) in the event of being unable to make the decision him- or herself.

> **health care proxy**
> A type of advance directive that designates a person to make treatment decisions for the patient in the event of being incapacitated.

> **durable power of attorney for health care decisions**
> A type of advance directive that grants full legal authority for treatment decisions to a designated person if the patient is incapacitated.

Living will: The living will specifies the treatment the patient desires to refuse (or accept) in the event of being unable to make the decision him- or herself. Examples include the refusal of intubation, CPR, feeding tubes, or being placed on a ventilator.

Health care proxy: The health care proxy designates a person to make treatment decisions for the patient in the event of being incapacitated.

Durable power of attorney for health care decisions: The durable power of attorney for health care decisions also designates a person to make treatment decisions but extends authority to a designated individual to make final treatment decisions, including cessation of treatment.

The DNR, or "do not resuscitate," is a physician's order dictating the limits to resuscitative effort. The DNR order is typically an expression of the advance directive.

Advance directives have a double edge. On one hand, the EMS professional must respect EMS duty to act and render appropriate emergency care. The scene of a cardiac arrest is not a convenient time to wait around for a copy of the patient's advance directive. On the other hand, the EMS professional must be respectful of resuscitation decisions that patients have expressed in their advance directives. Essential documentation elements of the advance directive include:

- Document the presence of the advance directive in the PCR. Follow, and document compliance with, local practice guidelines for management of the patient with an advance directive.
- If the patient or a family member states the patient has an advance directive but is unable to present the document, note this on the PCR and follow local practice guidelines.

ON TARGET EMS encounters proving to be a test of scene management and clinical skills will also be a test of your documentation skills.

- Document conversations with family members regarding EMS treatment and transport, using exact quotations as much as possible.
- Document direction received from medical control.
- As part of obtaining a patient's medical history, the EMS provider should routinely ask whether patients have an advance directive.

Summary: Return to Case Study

Three months after dispatch to the scene of a potential assault, you receive a subpoena calling you to testify in criminal trial. On the stand, you answer all questions, referring to your meticulously documented PCR. The family of the crime victim is relieved, and the prosecutor is pleased when the defense attorney states, "I have no further questions for this witness," and takes his seat.

The crime scene, the behavioral crisis, and issues associated with the rights of patients can be complex situations for the EMS professional. EMS encounters proving to be a test of scene management and clinical skills will also be a test of your documentation skills. By applying the principles presented in this text, out-of-the-ordinary patient encounters will have out-of-the-ordinary PCR documentation.

CHAPTER REVIEW

Review Questions

Please refer to Answers to Chapter Review Questions at the back of this book.

1. List the essential elements of crime scene documentation.

2. Describe the importance of documenting the neurological status of the crime scene patient.

3. List the essential documentation elements for the sexual assault patient.

4. List the essential documentation elements for the victim of child abuse.

5. List the essential documentation elements for the victim of elder abuse.

6. List the essential documentation elements for the patient with an order for physical restraints.

7. List the essential documentation elements for the suicidal patient.

8. List and describe the three types of advance directive.

Critical Thinking

Please refer to Answers to Critical Thinking Discussion Exercises at the back of this book.

1. Why is the patient's social history relevant to PCR documentation of the crime scene patient?

2. What types of patient encounters might an EMS professional have bias toward? Is there any correlation between personal bias and the manner in which certain EMS patient encounters are documented (for example, the patient with a behavioral emergency)?

Action Plan

1. Explore your attitudes toward patients experiencing a behavioral crisis.

2. Police departments are tremendous resources for EMS providers. If you are able, discuss EMS presence on crime scenes with a police officer you know. Discuss how police officers document crime scenes, and find out what is important from a law enforcement perspective.

3. Study elder abuse. Seek out resources to ensure you possess an adequate understanding of the signs and symptoms of elder abuse.

Practice Exercises—Role Play

Using role play, conduct an assessment to your level of licensure for each of the following patient encounters. Then, applying the principles presented in this chapter, document a PCR narrative, ensuring that all essential documentation elements are captured.

1. You are called to respond to the scene of a patient who has a gunshot wound to the left chest. A bloody handgun is found near the patient.

2. You are called to respond to the scene of a child, reported by his nanny, to have fallen backwards down a flight of stairs while playing with a neighbor. EMS is first on scene.

3. You are called to respond to the scene of a patient who has made an unsuccessful suicide attempt by cutting her wrists.

4. You are called to respond to the scene of a patient who states she has been sexually assaulted. EMS is first on scene.

Appendix A
Patient Care Report (PCR)

EMS Event Summary			
Dispatch Data			
Agency:	EMS Agency Number:	Incident #:	Unit #:
Type of Service:	Primary Unit Role:	Patient Care Report #:	
Pt. Record #	Trauma ID		
Response Mode:	ALS Nonemergency		ALS Emergency
BLS Emergency	BLS Nonemergency		Specialty Care Transport
Transport Mode:	ALS Nonemergency		ALS Emergency
BLS Emergency	BLS Nonemergency		Specialty Care Transport
Incident:			
Incident Address:	City	County	State	Zip
Response Delay: Y N Type:			
Transport Delay: Y N Type:			
Turnaround Delay: Y N Type:			
Odometer: Beginning On Scene Destination Ending			
Complaint Reported by Dispatch:			
Crew Member ID:			
1. _____ License Level _____ 2. _____ License Level _____			
Additional Crew Member: _____ License Level _____			
Additional Crew Member: _____ License Level _____			
Crew Member Role:			
Primary Caregiver Secondary Caregiver Driver			

Event Times:	Incident/Onset Time:
PSAP Call Time:	Unit Notified Time:
Unit En Route Time:	Unit on Scene Time:
Unit Left Scene Time:	Patient Destination Time:
Transfer of Patient Care Time:	Unit Back in Service Time:

Demographic Data:

Last Name	First Name	Middle

Mailing/Home Address: *(Street Address)*

City:	County:	State:	Zip:

SSN:	Gender:	Race:	Ethnicity:

Age:	Age Units:	Date of Birth:

Phone: (   )   DL # (State):

Financial Information:

Primary Method of Payment:
Medicare   Medicaid   Private Insurance   Self-Pay   WC
MVC/Accident Related:

Secondary Method of Payment:

Insurance Company Billing Priority: Primary   Secondary

Insurance Company Address
City: State: Zip Code:

Insurance Group/ID Name: Policy ID

Insured Information:
Last Name: First Name: Middle:
Relationship: Self   Spouse   Son/Daughter   Other

Work Related: Y   N
Occupational History: Patient Occupation:

Closest Relative/Guardian Information:
Last Name: First Name: Middle:
Street Address: City   State   Zip
Phone: Relationship:

Patient's Employer:
Address: City: State: Zip:

Patient's Work Telephone: (   )

Reason Information Not Obtained:

Assessment Summary

Scene Data:

Public Safety Agencies at Scene:	EMS Services at Scene:

Time Initial Responder on Scene:	Number of Patients at Scene:

Mass Casualty Incident Y N NA	Incident Location Type:

Incident Address:	City	County	State	Zip

Patient/Incident Disposition:

Canceled	Dead at Scene
No Patient Found	No Treatment Required
Patient Refused Care	Treated and Released
Treated/Transported	Treated/Care Transferred
Treated/Transported by Law Enforcement	Treated/Transported Private Vehicle

Situation Data:	Prior Aid:

Prior Aid By:	Outcome:

Possible Injury:

Chief Complaint: Primary	Secondary
Duration:	Duration:
Time Units:	Time Units:
Anatomic Location:	Anatomic Location:
Organ System:	Organ System:

EMS Diagnosis:
Primary:
Secondary:

Injury Summary:

Intent of Injury:	Unintentional	Intentional/Self

Cause of Injury:

Mechanism of Injury:

Burn	Penetrating	Blunt	Other

Vehicular Injury Indicators:

DOA	Fire	Ejection	> 1 Foot Space Intrusion	Rollover
Deformity:	Side	Steering Wheel	Dash	WS

Area of Vehicle Impact:
Center Left Rear Right Front Right Side Left Front Left Side Right Rear RO

Seat Row Location in Vehicle	Position of Patient in Seat

Use of Occupant Safety Equipment: None

Eye Protection	Lap Belt Protective Clothing Shoulder Belt Child Restraint
Helmet	Personal Flotation Device Protective Non-Clothing Gear

Airbag Deployment:	None	Side	Front	Other

Fall:	Y	N	Height:

Cardiac Arrest Summary:

Cardiac Arrest Y N After EMS Arrival Prior to EMS Arrival
Estimated Time of Arrest Prior to EMS:
Cardiac Arrest Etiology: Presumed Cardiac Drowning Electrocution Trauma Respiratory Other
Resuscitation Attempted: Attempted Defibrillation   Chest Compressions   Not Attempted/DNR   Attempted Ventilation   Not Attempted—Considered Futile
Arrest Witnessed By: First Monitored Rhythm:
Return of Spontaneous Circulation? Y   N
Reason CPR Discontinued: Time Resuscitation Discontinued:
Cardiac Arrest Narrative Comment:

Medical History Summary:

History Provided By: Patient   Family   Other
Sending Facility Medical Record #
Destination Medical Record #
Barriers to Assessment and Patient Care: None   Developmentally Impaired   Hearing Impaired   Physically Impaired   Speech /Language Impaired   Unconscious   Unattended/Unsupervised
Patient's Primary Practitioner: Last Name   First Name   Middle Name
Advance Directives: DNR Form   State/EMS DNR Form   Living Will   Family/Guardian Request (No Documentation)   Other   None
Allergies: Medication: Environmental/Food:
Medical/Surgical History: Pertinent Family History:
Social History Note: Stressors: Nicotine Use: Y N Packs/Day/Years: Support Structure: Abuse Potential:

Current Medications:			
Medication	**Dose**	**Unit**	**Route**
1.			
2.			

Presence of Emergency Information Form: Yes   No   Not Available   NA

Alcohol/Drug Use Indicators: None/NA   Smell of Alcohol on Breath
  Patient Admits to Drug Use   Patient Admits to Alcohol Use   Other

Pregnancy: Y   N   Not Applicable   Not Available   Not Known

Assessment/Vital Signs	TIME	BP	Heart Rate	RESP	02 SAT	TEMP	EKG	PAIN	GLUCOSE	ETCO2
VS Obtained Prior to EMS: _____ Obtained By:		Method:	Rhythm:	Effort:						
Glasgow Coma Scale Eyes – Verbal – Motor – TOTAL: _____		Method:	Rhythm:	Effort:						
Level of Responsiveness: Alert Verbal Unresponsive Painful Stimuli		Method:	Rhythm:	Effort:						
Oriented: Person Place Time Event		Method:	Rhythm:	Effort:						

Stroke Scale:
Cincinnati Stroke Scale Non-conclusive   LA Stroke Scale Negative
Cincinnati Stroke Scale Negative   LA Stroke Scale Positive
Cincinnati Stroke Scale Positive   LA Stroke Scale Non-conclusive

Apgar:
Color —	0 = Blue/Pale	1 = Body Pink/Extremities Blue	2 = Pink
HR —	0 = Absent	1 = <100	2 = >100
Muscle Tone —	0 = Limp	1 = Some Flexion	2 = Good Flexion
Reflect Irritability —	0 = No Response	1 = Some Motion	2 = Cry
Respiratory Effort —	0 = Absent	1 = Weak Cry	2 = Strong Cry

TOTAL: _____

Trauma Score:

Assessment Note:

Trauma Assessment:

Injury Skin:		Soft Tissue Swelling/Bruising
Bleeding Controlled	Burn	Dislocation Fracture
Puncture/Stab	Amputation	Bleeding Uncontrolled
Crush	Gunshot	Pain w/o Swelling/Bruising

Injury Face/Head:		Soft Tissue Swelling/Bruising
Bleeding Controlled	Burn	Dislocation Fracture
Puncture/Stab	Amputation	Bleeding Uncontrolled
Crush	Gunshot	Pain w/o Swelling/Bruising

Injury Neck:		Soft Tissue Swelling/Bruising
Bleeding Controlled	Burn	Dislocation Fracture
Puncture/Stab	Amputation	Bleeding Uncontrolled
Crush	Gunshot	Pain w/o Swelling/Bruising

Injury Thorax:		Soft Tissue Swelling/Bruising
Bleeding Controlled	Burn	Dislocation Fracture
Puncture/Stab	Amputation	Bleeding Uncontrolled
Crush	Gunshot	Pain w/o Swelling/Bruising

Injury Abdomen:		Soft Tissue Swelling/Bruising
Bleeding Controlled	Burn	Dislocation Fracture
Puncture/Stab	Amputation	Bleeding Uncontrolled
Crush	Gunshot	Pain w/o Swelling/Bruising

Injury Spine:		Soft Tissue Swelling/Bruising
Bleeding Controlled	Burn	Dislocation Fracture
Puncture/Stab	Amputation	Bleeding Uncontrolled
Crush	Gunshot	Pain w/o Swelling/Bruising

Injury Upper Extremities:		Soft Tissue Swelling/Bruising
Bleeding Controlled	Burn	Dislocation Fracture
Puncture/Stab	Amputation	Bleeding Uncontrolled
Crush	Gunshot	Pain w/o Swelling/Bruising

Injury Pelvis:		Soft Tissue Swelling/Bruising
Bleeding Controlled	Burn	Dislocation Fracture
Puncture/Stab	Amputation	Bleeding Uncontrolled
Crush	Gunshot	Pain w/o Swelling/Bruising

Injury Lower Extremities:		Soft Tissue Swelling/Bruising
Bleeding Controlled	Burn	Dislocation Fracture
Puncture/Stab	Amputation	Bleeding Uncontrolled
Crush	Gunshot	Pain w/o Swelling/Bruising

Trauma Assessment Note:

Focused Physical Examination:

Estimated Body Weight: Time of Assessment:

General Appearance
Grooming: Appropriate   Not Appropriate   Describe:

Gestures/Body Language: Appropriate   Describe: _____

Positioning:

Mental Status Evaluation:
Speech: Normal   Slurring   Normal/Coherent   Non-Appropriate   Flight of Ideas

Emotional State: Depression   Elation   Anger   Other: _____

Thought: Anxious   Delusions   Hallucinations   Other: _____

Cognitive Abilities: Attention Intact   Memory Intact   Deficits _____

Neurological
Seizure Activity: Onset: _____   Past History   Y   N

Generalized: Tonic-Clonic   Absence   Pseudoseizure

Partial: Simple Partial   Complex Partial

Motor: Weakness –   RUE   RLE   LUE   LLE   Generalized

Gait: Normal   Abnormal   Describe: _____

Sensory: Pain Status: _____:10   Location: _____

             EMS Field Cranial Nerve Testing:

I: Smell   VIII: Hearing

II: Vision   IX/X: Swallow Ability

III/IV/VI: Eye and Eye Lid Movement   XI: Shoulder Shrug

V: Facial Muscle Tone   XII: Tongue Movement

VII: Raise Eyebrows/Show Teeth

HEENT -
Head: Deformity   Tenderness   Trauma   No Findings

Eyes: Symmetrical   Asymmetrical   Drainage:

Pupils Left: 2 mm 3 mm 4 mm 5 mm 6 mm 7 mm Fixed Reactive Non-Reactive

Eyes Right: 2 mm 3 mm 4 mm 5 mm 6 mm 7 mm Fixed Reactive Non-Reactive

Ears: Hearing Intact   Y   N   Drainage: Y   N Describe:_____

Nose: Nasal Flaring Deviation Deformity Tenderness Drainage Describe: _____

Mouth/Throat/Neck: Mucous Membranes Moist: Y   N Teeth Intact: Y   N

Trachea: Midline   Left Deviation   Right Deviation   JVD: Left   Right

Respiratory Status:
                       Respiratory Pattern

Normal Bradypnea Tachypnea Hyperventilation Cheyne-Stokes Kussmaul

Apneustic
Respiratory Effort

Symmetrical   Retractions   Increased Effort   Use of Accessory Muscles
Cough
None   Non-Productive   Productive   Describe: _____

Breath Sounds
Normal

Decreased BS: Left   Right   Rhonchi: Left   Right
Crackles: Left   Right   Wheezing: Left   Right

Cardiovascular Status:
                                    Skin Assessment

Warm   Dry   Pale   Clammy   Cold   Cyanotic   Mottled   Jaundiced Skin
Nail Beds
Normal   Cyanotic   Clubbing
PMI
          Location:_____   Thrills   Bruits
Heart Tones
S1 S2 S3 S4   Friction Rub   Murmur
Peripheral Edema
4+ = Very Deep/2–6 minutes   3+ = Noticeably Deep – 1 minute   2+ = Disappears in 10–20
seconds   1+ = Slight Pitting—Disappears Quickly
RLE ___   LLE _____   RUE _____ LUE _____   Abdomen_____   Other: _____
Pulses
4 = Bounding   3 = Full   2 = Expected   1 = Diminished   0 = Absent
Carotid: _____   Brachial: _____   Radial: _____   Femoral: _____
Popliteal: _____   Dorsalis pedis: _____   Posterior tibial: _____

GI/GU Status:
Abdominal Appearance: Flat   Round   Distended   Symmetrical   Asymmetrical
Bowel Sounds: Present   Absent   Diminished   Increased
Abdomen LU Assessment—   Guarding Tenderness Normal   Distension   Mass
Abdomen LL Assessment—   Guarding Tenderness Normal   Distension   Mass
Abdomen RU Assessment—   Guarding Tenderness Normal   Distension   Mass
Abdomen RL Assessment—   Guarding Tenderness Normal   Distension   Mass
Urinary Status: Dysuria Hematuria Polyuria Hesitancy (-)Stream Incontinence
Female: Discharge   STD: _____
Male: Swelling   Torsion   Lesions   Discharge STD:_____

Musculoskeletal:
Cervical: Normal   Tender/Para-Spinous   Pain to ROM   Tender Spinous Process
Thoracic: Normal   Tender/Para-Spinous   Pain to ROM   Tender Spinous Process
Lumbar: Normal   Tender/Para-Spinous   Pain to ROM   Tender Spinous Process
RUE: Tenderness/Pain   Redness/Warmth Weakness   Deformity   Decreased ROM
RLE: Tenderness/Pain   Redness/Warmth Weakness   Deformity   Decreased ROM
LUE: Tenderness/Pain   Redness/Warmth Weakness   Deformity   Decreased ROM
LLE: Tenderness/Pain   Redness/Warmth Weakness   Deformity   Decreased ROM

OB:

Gravida: _____ Para: _____ Gestational Age: _____ EDC: _____
Vaginal Bleeding ROM: Y   N   Unknown   Time: _____ Color: _____
Contractions: Onset Time: _____ Timing: Every _____ Duration: _____

Assessment Note:

EMS Treatment

BLS Interventions:

Interventions Performed Prior to EMS Care:

Intervention Authorization: On-Scene   Written Order   On-Line   Protocol

Airway Management:
Airway Status: Clear   Obstructed   Compromised
Suction: Y   N   Device   Pre-oxygenated Y   N
Time: Device: Notes: _____
Oxygen Therapy: Device: Time: Flow Rate:
Pre-Oxygen Saturation: Post-Oxygen Saturation:

Immobilization:

Spinal	Extremity
Time:	Time:
Device:	Device:
Pre-Neuro Status:	Pre-Neuro Status:
Post-Neuro Status:	Post-Neuro Status:
Destination Status:	Destination Status:

Intervention Narrative Comment:

Airway Interventions

Procedure	Time	Equipment Used/Size	Attempts	Successful	Response	Staff ID
				Y N		
				Y N		
				Y N		

Airway Grade:

Intubation Adjuncts:

Complications: Rescue Airway:

ET Tube Confirmation:

Color CO2 Detector   Auscultation of Bilateral Breath Sounds

Esophageal Bulb Aspiration	Digital CO2 Confirmation
Waveform CO2 Confirmation	Negative Auscultation over Epigastrium
Visualization of Tube–Cords	
Cuff Inflated: Y N	N/A

Destination Tube Confirmation:

Color CO2 Detector	Auscultation of Bilateral Breath Sounds
Esophageal Bulb Aspiration	Digital CO2 Confirmation
Waveform CO2 Confirmation	Negative Auscultation over Epigastrium
Visualization of Tube–Cords	

Intervention Narrative Comment:

Cardiac Monitoring/Interventions

EKG Lead Monitored: Time Initiated:

12 EKG: EKG Time:
ST Elevation Leads:
ST Depression Leads:
EKG Changes: Time:

Defibrillation: Monophasic   Biphasic

Time	Rhythm	Energy Level	Secondary Rhythm	Defibrillation/ Cardioversion	Staff ID

Total Number of Shocks Delivered:

Transcutaneous Pacing:
Rate: Energy Level: Capture: Time:

Intervention Narrative Comment:

Interventions/Procedures—IV Therapy

Procedure Performed Prior to EMS Care: Y   N   Agency:
Site Assessment: Infusion Rate Verification:

Allergies Checked: Y   N   None

Site Evaluation:
Site Prep:

Time	Site	Catheter	Set	Attempts	Fluid/Rate	Staff ID

Successful:	Y	N	Complications:		
Total Attempts:	1	2	3	MCP Notified:	Y N
Post Intervention Site Assessment:					

Maintenance IV:

Site Assessment: _____ Fluid: _____ Additives: _____

Rate: Verified in Medical Records: Y   N

Site Evaluation at Destination:

Intervention Narrative Comment:

Interventions/Medication Administration

Medications Administered Prior to EMS Care:

Medication Authorization:　　　Standing Orders　　　Written Order　　　On-Line　　Protocol

Allergies Checked/Source:

Medications

Time	Medication	Dose/Units	Route	Response	Staff ID

Intervention Narrative Comment:

Interventions/Safety:
PPE Used:
Secured to Ambulance Stretcher:

Patient Moved to Ambulance:

Carry	Assisted/Ambulated	Other
Stretcher	Stair chair	

Post Movement Assessment:

Secured to Ambulance Stretcher:

Patient Moved from Ambulance:

Carry	Assisted/Ambulated	Other
Stretcher	Stair chair	

Post Movement Assessment:

Patient Belongings:	Secured for Transport:	Y	N

Intervention Narrative Comment:

Changein Patient Status:

Change:	Time:	Intervention:
Change:	Time:	Intervention:
Change:	Time:	Intervention:

Status Change Note:

Affirmation

Medical Necessity of Emergency Medical Service:

EMS Diagnosis:
Condition Code:

Past Medical History Contributing to This EMS Incident:

Bed-Confinement Status:
At the time of the transport was the patient:

1. Unable to get up from bed without assistance?	Y	N
2. Unable to ambulate; and	Y	N
3. Unable to sit in a chair or wheelchair?	Y	N

Reason for Above Bed-Confinement:
Contributing Diagnosis:

Inter-Facility Transports:
Service(s) Not Available at Sending Facility:
Immediate Transport Needed:
Reason:
Time Requirement:
Equipment Required:

Medical Necessity for Transport to a Medical Facility of Greater Distance:
Facility Bypassed:
Services or MD Unavailable:

Return to Facility Post Procedure:	Procedure: _____	Date: _____

Patient/Family Preference Only:	Y	N

Other:

Sending MD:	Receiving MD:

PCS Attached:	Y	N	Not Applicable	
ABN Signed:	Y	N	Not Applicable/Emergency Patient	

Transport/Loaded Miles:

Medical Necessity Not Established:	Reason:

Medical Necessity Statement:

Transfer of Care:

Receiving Facility:

Patient/Incident Disposition:	
Canceled	Dead at Scene
No Patient Found	No Treatment Required
Patient Refused Care	Treated and Released
Treated/Transported	Treated/Care Transferred
Treated/Transported by Law Enforcement	Treated/Transported Private Vehicle

Transport Mode from Scene:

Condition of Patient at Destination:

Reason for Choosing Destination Facility:

Destination Type:

Radio/Phone Report to Destination Facility:	
Time:	Person:

Transfer of Care:	I have received an appropriate Transfer of Care:
Time:	Title:
Signature:	Medical Records Received:

EMS Event Summary:

EMS Diagnosis:
Scene Summary:
Review of Systems:
Intervention Summary:
Safety Summary:
Disposition Summary:

Signatures:	
Patient Signature:	Date/Time:
Belongings Received (Type):	
Staff Signatures:	

Signature:	License #
Signature:	License #
Peer Reviewed: Y N	

Appendix B
Standard Charting Abbreviations

Patient Information/Categories

Asian	A
Black	B
Chief complaint	CC
Complains of	c/o
Current health status	CHS
Date of birth	DOB
Differential diagnosis	DD
Estimated date of confinement	EDC
Family history	FH
Female	♀
Hispanic	H
History	Hx
History and physical	H&P
History of present illness	HPI
Impression	IMP
Male	♂
Medications	Med
Newborn	NB
Past history	PH
Patient	Pt
Physical exam	PE
Private medical doctor	PMD
Signs and symptoms	S/S
Vital signs	VS
Weight	Wt
White	W
Year-old	y/o

Body Systems

Abdomen	Abd
Cardiovascular	CV
Central nervous system	CNS
Ear, nose, and throat	ENT
Gastrointestinal	GI
Genitourinary	GU
Gynecological	GYN
Head, eyes, ears, nose, and throat	HEENT
Musculoskeletal	M/S
Obstetrical	OB
Peripheral nervous system	PNS
Respiratory	Resp

Common Complaints

Abdominal pain	abd pn
Chest pain	CP
Dyspnea on exertion	DOE
Fever of unknown origin	FUO
Gunshot wound	GSW
Headache	H/A
Lower back pain	LBP
Nausea/vomiting	n/v
No apparent distress	NAD
Pain	pn
Shortness of breath	SOB
Substernal chest pain	sscp

Diagnoses

Abdominal aortic aneurysm	AAA
Abortion	Ab
Acute myocardial infarction	AMI
Adult respiratory distress syndrome	ARDS
Alcohol	ETOH
Atherosclerotic heart disease	ASHD
Chronic obstructive pulmonary disease	COPD
Chronic renal failure	CRF
Congestive heart failure	CHF
Coronary artery bypass graft	CABG
Coronary artery disease	CAD
Cystic fibrosis	CF
Dead on arrival	DOA
Deep vein thrombosis	DVT
Delirium tremens	DTs
Diabetes mellitus	DM
Dilation and curettage	D&C
Duodenal ulcer	DU
End-stage renal failure	ESRF
Epstein–Barr virus	EBV

Foreign body obstruction	FBO	Nitroglycerin	NTG
Hepatitis B virus	HBV	Nonsteroidal anti-inflammatory agent	NSAID
Hiatal hernia	HH	Normal saline	NS
Hypertension	HTN	Penicillin	PCN
Infectious disease	ID	Phenobarbital	PB
Inferior wall myocardial infarction	IWMI	Potassium	K^+
Insulin-dependent diabetes mellitus	IDDM	Sodium bicarbonate	$NaHCO_3$
Intracranial pressure	ICP	Sodium chloride	NaCl
Mass casualty incident	MCI	Tylenol	APAP
Mitral valve prolapse	MVP		
Motor vehicle crash	MVC	*Anatomy/Landmarks*	
Multiple sclerosis	MS	Abdomen	Abd
Non-insulin-dependent diabetes mellitus	NIDDM	Antecubital	AC
Organic brain syndrome	OBS	Anterior axillary line	AAL
Otitis media	OM	Anterior cruciate ligament	ACL
Overdose	OD	Anterior–posterior	A/P
Paroxysmal nocturnal dyspnea	PND	Dorsalis pedis (pulse)	DP
Pelvic inflammatory disease	PID	Gallbladder	GB
Peptic ulcer disease	PUD	Intercostal space	ICS
Pregnancies/births *(gravida/para)*	G/P	Lateral collateral ligament	LCL
Pregnancy-induced hypertension	PIH	Left lower lobe	LLL
Pulmonary embolism	PE	Left lower quadrant	LLQ
Rheumatic heart disease	RHD	Left upper lobe	LUL
Sexually transmitted disease	STD	Left upper quadrant	LUQ
Transient ischemic attack	TIA	Left ventricle	LV
Tuberculosis	TB	Liver, spleen, and kidneys	LSK
Upper respiratory infection	URI	Lymph node	LN
Urinary tract infection	UTI	Midaxillary line	MAL
Wolff–Parkinson–White syndrome (disease)	WPW	Posterior axillary line	PAL
		Right lower lobe	RLL
Medications		Right lower quadrant	RLQ
Angiotensin-converting enzyme	ACE	Right middle lobe	RML
Aspirin	ASA	Right upper lobe	RUL
Bicarbonate	HCO_3	Right upper quadrant	RUQ
Birth control pills	BCP	Temporomandibular joint	TMJ
Calcium	Ca^{++}	Tympanic membrane	TM
Calcium channel blocker	CCB		
Calcium chloride	$CaCl_2$	*Physical Exam/Findings*	
Chloride	Cl^-	Arterial blood gas	ABG
Digoxin	Dig	Bilateral breath sounds	BBS
Dilantin (phenytoin sodium)	DPH	Blood sugar	BS
Diphendydramine	DPHM	Breath sounds	BS
Diphtheria–Pertussis–Tetanus	DPT	Cerebrospinal fluid	CSF
Hydrochlorothiazide	HCTZ	Chest X-ray	CXR
Lactated Ringer's, Ringer's Lactate	LR, RL	Complete blood count	CBC
Magnesium sulfate	Mg^{++}	Computerized tomography	CT
Morphine sulfate	MS	Conscious, alert, and oriented	CAO
		Costovertebral angle	CVA

Deep tendon reflexes	DTR	Fahrenheit	F°
Dorsalis pedis (pulse)	DP	Immediately	stat
Electrocardiogram	EKG, ECG	Increased	↑
Electroencephalogram	EEG	Inferior	inf.
Expiratory	Exp	Left	Ⓛ
Extraocular movements (intact)	EOMI	Less than	<
Fetal heart tones	FHT	Moderate	mod.
Full range of motion	FROM	More than	>
Full-term normal delivery	FTND	Negative	—
Heart rate	HR	No, not, none	Ø
Heart sounds	HS	Not applicable	n/a
Hemoglobin	Hgb	Number	No or #
Inspiratory	Insp	Occasional	occ
Jugular venous distention	JVD	Pack years	pk/yrs, p/y
Laceration	lac	Per	/
Level of consciousness	LOC	Positive	+
Moves all extremities (well)	MAEW	Posterior	post.
Nontender	NT	Postoperative	PO
Normal range of motion	NROM	Prior to arrival	PTA
Palpation	Palp	Radiates to	→
Passive range of motion	PROM	Right	®
Point of maximal impulse	PMI	Rule out	R/O
Posterior tibial (pulse)	PT	Secondary to	2°
Pulse	P	Superior	sup.
Pupils equal and reactive to light	PEARL	Times (for 3 hours)	×(× 3h)
Pupils equal, round, reactive to light and accommodation	PERRLA	Unequal	≠
		Warm and dry	W/D
Range of motion	ROM	While awake	WA
Respirations	R	With (*cum*)	c̄
Temperature	T	Within normal limits (or we never looked)	WNL
Unconscious	unc	Without (*sine*)	s̄
Urinary incontinence	UI	Zero	0

Miscellaneous Descriptors

Treatments/Dispositions

After (post-)	p̄	Advanced cardiac life support	ACLS
After eating	pc	Advanced life support	ALS
Alert and oriented	A/O	Against medical advice	AMA
Anterior	ant.	Automated external defibrillator	AED
Approximate	≈	Bag–valve mask	BVM
As needed	prn	Basic life support	BLS
Before (ante-)	ā	Cardiopulmonary resuscitation	CPR
Before eating (*ante cibum*, before meal)	a.c.	Continuous positive airway pressure	CPAP
Body surface area (%)	BSA	Do not resuscitate	DNR
Celsius	C°	Endotracheal tube	ETT
Change	Δ	Estimated time of arrival	ETA
Decreased	↓	External cardiac pacing	ECP
Equal	=	Intermittent positive-pressure ventilation	IPPV

Long spine board	LSB	Hydrogen-ion concentration	pH
Nasal cannula	NC	Intracardiac	IC
Nasogastric	NG	Intramuscular	IM
Nasopharyngeal airway	NPA	Intraosseous	IO
No transport—refusal	NTR	Intravenous	IV
Nonrebreather mask	NRM	Intravenous push	IVP
Nothing by mouth	NPO	Joules	j
Oropharyngeal airway	OPA	Keep vein open	KVO
Oxygen	O_2	Kilogram	kg
Per square inch	psi	Liter	L
Physical therapy	PT	Liters per minute	LPM, L/min
Positive end-expiratory pressure	PEEP	Microgram	mcg
Short spine board	SSB	Milliequivalent	mEq
Therapy	Rx	Milligram	mg
Treatment	Tx	Milliliter	ml
Turned over to	TOT	Millimeter	mm
Verbal order	VO	Millimeters of mercury	mmHg
		Minute	min

Medication Administration/Metrics

Centimeter	cm	Orally	po
Cubic centimeter	cc	Subcutaneous	SC, SQ
Deciliter	dL	Sublingual	SL
Drop(s)	gtt(s)	To keep open	TKO
Drops per minute	gtts/min		

Cardiology

Every	q	Atrial fibrillation	AF
Grain	gr	Ventricular fibrillation	VF
Gram	g, gm	Ventricular tachycardia	VT
Hour	h or hr		

Answers to
Case Study Questions

The purpose of the case study questions is to encourage identification and evaluation of your perceptions, attitudes, and practice habits in documentation. These questions are designed to prompt you to think about "what I would do in this situation" and to promote application of chapter content to EMS practice.

Chapter 1, page 1

1. What is your definition of EMS documentation?

   Discussion: Definition defines purpose. Your definition of EMS documentation reveals why you document. If your purpose for documenting is limited to the task of simply recording clinical care, the importance of PCR documentation will be limited in your EMS practice. A limited definition limits the capacity for professional growth.

2. How does documentation match up to other EMS skills such as spinal immobilization and airway management?

   Discussion: If you consider documentation a skill, you then have the capacity to improve upon that skill in the same manner, for example, as spinal immobilization. If you do not have proficiency in spinal immobilization, a patient may never walk again, so great effort goes into developing that skill. If documentation is viewed as a skill that must be developed in order to attain proficiency, the EMS professional will have taken the first step toward documentation proficiency.

3. What is the purpose of the Patient Care Report?

   Discussion: Your understanding of the purposes of documentation will assist you in expanding your horizons beyond simply filling in blanks to documenting your professional activities.

Chapter 2, page 13

Evaluate the Patient Care Report in Figure 2-2.

1. Based only on the documentation, how would you evaluate the patient care?

   Discussion: Patient care is evaluated in two ways: first, by direct observation in real time by a field training officer, supervisor, or medical director (usually against practice standards); second, by what has been documented about the care in the PCR. In EMS, quality of care is almost always evaluated by what the PCR states about the patient care. Rarely in your career will your practice be evaluated by the direct observation of a training officer or medical director.

2. Describe the link between quality of care and quality of documentation.

   Discussion: You think you performed well on a call, and your partner agrees. Upon arrival at the ED the patient is nicely packaged for the staff waiting to take over care, and from all appearances the patient received quality care. Within minutes, the good feelings are forgotten and quality will be evaluated in light of the care that was documented. PCR documentation is the link between perceived quality and actual quality.

3. How does the documentation in the PCR in Figure 2-2 reflect on the EMS provider?

   Discussion: A legible and complete PCR, documenting care in line with practice standards, reflects positively on the EMS provider. Likewise, a sloppy and incomplete PCR communicates sloppy and incomplete patient care and reflects negatively on the EMS professional.

4. The EMS provider in the case study is in trouble. Could it have been avoided?

   Discussion: The care given by the EMS provider in the case study was textbook perfect. However, the PCR failed to reflect textbook perfect care. The unfortunate events in the case study (and in real-life EMS practice) can be avoided by taking the time to meticulously document textbook-perfect care.

Chapter 3, page 31

1. What made the EMS services in the case study nonpayable?

   Discussion: Accuracy in documentation is linked to accurate reimbursement. PCR documentation that fails to capture accurately the EMS services will make it difficult for EMS billing specialists to bill properly for services.

2. What role does PCR documentation have in the financing of EMS?

   Discussion: The PCR is a financial instrument by which EMS services are evaluated for payment. Although the EMS professional must never document specifically for payment purposes, there should be an awareness of the financial functions of documentation.

Chapter 4, page 45

1. If your patient care were put to the test of a negligence lawsuit, do you think it would it affect your relationships with your peers?

   Discussion: EMS professionals that have gone through litigation attest to how difficult an experience it can be and how it affects just about everything. Attaining proficiency in documentation is an excellent strategy for avoiding this situation.

2. What is the difference between documentation and patient care?

   Discussion: Documentation and patient care have traditionally been thought of as very separate from each other. Documentation is something done after the EMS encounter that simply records patient care. One of the main purposes of this chapter is to communicate that documentation and patient care are inseparable. PCR documentation, as a professional activity, is patient care for which the EMS professional can be held legally accountable.

3. Is it more likely for someone who is angry or grieving to file a lawsuit? Why or why not, and how does this tie to documentation?

   Discussion: An angry or grieving person often feels the need to strike back, obtain justice, or punish a health care professional when expectations are unmet. An accurate, complete PCR is the EMS professional's protection from the legal vulnerabilities that can result from unmet expectations.

Chapter 5, page 63

1. Do you think there is anything wrong with the organization's approach to the "documentation problem"? Why or why not?

   Discussion: The case study illustrates one of the reasons documentation is controversial and problematic within many EMS systems—the varied financial, operational, and quality interests within the organization.

2. What will be necessary for this initiative to be successful?

   Discussion: In order for this effort to be successful, it will need to be understood that PCR documentation is linked to the entire organization. Each department, while having different perspectives on documentation, is dependent upon EMS staff becoming proficient in documentation.

3. Do you think your perspective on EMS documentation would change if you were placed in a QI role?

   Discussion: Being involved in a quality improvement role can be an eye-opening experience that reveals the weaknesses and strengths of clinical care within the organization. The PCR is usually the instrument of quality review. Therefore, conclusions regarding quality are often based solely on documentation.

Chapter 6, page 77

1. What are the advantages of using an electronic PCR? What are the disadvantages?

   Discussion: Advantages: The electronic PCR eliminates the legibility problem in documentation. Other strengths include "forcing" documentation with mandatory data entry fields, drop-down menus, and other clinical tools. Many ePCRs will assign a condition code and level of service to the EMS encounter, greater expediting the billing process. Disadvantages: The ePCR is only as good as the software, and the EMS provider must be careful not to point and click his or her way to lackluster documentation. When data entry choices fail to capture the EMS encounter adequately, greater effort will be required in narrative documentation.

2. Why is the electronic PCR essential to data collection?

Discussion: Data collection cannot be manually tabulated for the thousands of EMS encounters each day. Datasets and data elements have assigned numerical values, allowing electronic data extraction and collection of usable data for subsequent evaluation.

Chapter 7, page 105

1. What is the purpose of the narrative in EMS documentation?

Discussion: Understanding the purpose of the narrative is essential to effective narrative documentation. The narrative ties together the crucial elements of the EMS event, affirms medical necessity for EMS services, educates the reader, explains exceptions, and reinforces that EMS services were safely provided.

2. What are the dangers of summarizing the entire EMS encounter in the narrative section?

Discussion: If the narrative is used to provide a play-by-play account of the entire EMS encounter, there will always be the danger of leaving out critical information. If the narrative is used to summarize everything, then everything must be in the summary.

Chapter 8, page 119

1. Why is medical necessity often frustrating and controversial?

Discussion: Many in EMS have lacked understanding of the principles and regulations governing medical necessity and how medical necessity is established.

2. Is medical necessity documentation important to EMS? Why or why not?

Discussion: Medical necessity documentation is essential for appropriate reimbursement and the funding of EMS.

3. Is determining and evaluating medical necessity your job, or is it the job of the billing department staff?

Discussion: Determining and evaluating medical necessity is the responsibility of EMS billing staff. However, they are often unable to do their jobs because the information needed to evaluate medical necessity has not been obtained and documented. It is the EMS professional's responsibility to obtain and document the information relating to medical necessity in order for the billing staff to evaluate medical necessity and appropriately bill for the EMS services.

Chapter 9, page 141

1. Do you see any issues in the case study with the EMS provider's approach to documentation? If so, what are the issues and what may be the solution?

Discussion: When you're busy, keeping up with documentation can be a challenge. The issue in the case study could simply be a busy shift, with not enough time in between calls to finish documentation. Or it could indicate a more serious problem in how one manages his or her EMS practice.

2. Describe your current documentation practices. Do you complete each PCR at the end of each EMS encounter, or do you wait until the end of the shift? Do you keep up with documentation when call volume is high?

Discussion: Take a step back and evaluate how you manage documentation. If you habitually leave documentation to the end of the shift, and complete PCRs hours after the EMS encounter, this is problematic. Crucial information can be lost, and there is the danger of the hospital record being incomplete and inconsistent with the "final version" of the PCR.

Chapter 10, page 149

1. How would you evaluate the manner in which this patient refusal was managed?

Discussion: As you evaluate the management of the patient in the Case Study, three clues are found in the words appears, probably, *and* quickly. *The patient refusing EMS requires the EMS professional's time, undivided attention, and thorough evaluation in order to meet his or her legal and professional obligations.*

2. Would you have handled this patient differently? If so, how would you have managed the patient's refusal?

Discussion: Evaluating another's practices in respect to patient refusals can shed light on how you manage patients refusing EMS services. In this chapter we will discuss the proper management and documentation of patient refusals.

Chapter 11, page 163

1. How would you document the incident in the case study?
Discussion: Consider how you would document this incident if you found yourself in this situation. Write a short narrative in the space provided. After you have completed the chapter, review the narrative and determine if you would document the incident differently.

2. Would you use the PCR or another form or both?
Discussion: It is easy to take a "one form fits all" approach in documentation, and the result is that inappropriate information can be documented in the PCR. In this chapter we will discuss the appropriate use of the incident report and differentiate what should be documented using the incident report from the patient care documented in the PCR.

Chapter 12, page 173

1. Did the radio report provide the appropriate information to the receiving hospital? Why or why not?
Discussion: The radio report must give the receiving hospital only the information needed to prepare for the patient.

2. Was the information provided in the verbal report to the RN at the hospital appropriate?
Discussion: First, it is dangerous to assume the RN receiving the patient at the hospital was the one who received the radio report. The report to the RN at the receiving facility must provide all necessary information regarding the patient's presenting illness/injury, past medical history, and EMS treatment in order for a proper transfer of care to occur.

3. Evaluate the manner in which the nurse was advised of the "additional medication." Was this appropriate?
Discussion: Verbal reports must respect the patient's privacy and guard the confidentiality of protected health information (HIPAA regulations). The EMS professional also must be careful to assess whether the RN receiving the report is giving proper attention to the report.

Chapter 13, page 185

1. Are complex patient encounters any different from other "everyday" EMS encounters? Why or why not?
Discussion: Complex patient encounters are different in respect to the added complexity of scene management, integration with law enforcement, and issues relating to patient's rights.

2. If you were advised by a law enforcement officer that a patient encounter was going to involve a criminal investigation, would it change your approach to PCR documentation? Why or why not?
Discussion: There can be a tendency to take a different approach to documenting "the big ones" that are perceived to be on the fast track to the courtroom. EMS encounters considered to be mundane often receive less attention in documentation. Instead, the EMS professional must have one approach to all EMS encounters using a methodical, structured approach to PCR documentation.

Answers to
Chapter Review Questions

Chapter 1, page 1

1. List the three major challenges facing EMS today. Describe how the EMS professional can make an impact in meeting these challenges through the PCR.
   - *Establishing EMS as a profession*
   - *The increasing legislative, regulatory, and legal involvement in EMS*
   - *Reimbursement*

   Becoming proficient in documentation reflects positively on the EMS professional, reduces the risk of litigation, demonstrates compliance with EMS statutes and regulations, and assists the EMS organization in obtaining appropriate reimbursement.

2. Describe how PCR documentation reflects upon you as an EMS professional.

   The EMS professional practices before a large audience. PCRs are reviewed for quality, administrative, reimbursement, and legal purposes. Quality documentation leaves a positive impression of the patient care provided by the EMS professional on the reviewer.

3. Why is documentation an EMS skill? Describe the importance of attaining proficiency in documentation skill to your EMS career.

   Documentation is a skill because it requires training and practice to attain proficiency. Documentation permanently records all the other skills used to perform your patient care and authenticates whether or not these skills were performed correctly.

4. Describe how developing proficiency in documentation protects the EMS professional's career.

   Attaining proficiency in documentation protects the EMS professional's career by authenticating that patient care was provided within the boundaries of practice standards and greatly reduces the risks associated with litigation.

5. Evaluate the PCR in Figure 1-2.
   - What are your impressions of this documentation?
   - How would you describe the patient care based upon what has been documented?
   - How does this PCR represent EMS?

Evaluation points to consider
   - *Is documentation of the chief complaint descriptive enough?*
   - *Are allergies documented?*
   - *Is documentation of how the patient was found descriptive enough?*
   - *Times of vital signs are missing.*
   - *Only one set of vital signs taken for a patient with chest pain.*
   - *Nitroglycerin administration times not documented.*
   - *The patient's response to nitroglycerin not documented.*
   - *No examination documented in the narrative.*
   - *Interventions not documented in the narrative.*
   - *EMS encounter not summarized in the narrative.*
   - *No patient signature obtained (or reason signature was not obtained).*

Chapter 2, page 13

1. Why is it important for the EMS profession to be clearly defined?

   Defining a profession provides a template, guiding the professional behaviors and practice of the professional. As EMS becomes more clearly defined as a profession, its attributes can be displayed in everyday EMS practice.

2. List attributes of the emerging EMS profession.

National identity: A profession has an established national identity with a unified national association that leads and speaks for members of the profession.

Defined leadership: A profession has well-defined leadership from within the profession. For the EMS profession to advance, leadership must come from a single professional association that leads and speaks for members of the profession.

Specialized EMS knowledge and education: A profession has specialized knowledge and education. EMS knowledge and education is unique to health care and provides a platform to display the uniqueness of the profession through daily EMS practice.

Autonomy: A profession is autonomous when professional behaviors and practice are directed from within the profession.

High ethical standards: A profession demands high ethical standards of its membership. As EMS emerges as a profession, members will have ownership for establishing and enforcing high ethical standards.

3. What was the focus of documentation in the 1960s and 1970s?

In the 1960s documentation usually centered on the basic elements of transportation: name, address, pickup location, hospital, mileage, problem, and charge amount. In the 1970s documentation reflected the emerging technician because the first trip sheets began recording basic clinical care: chief complaint, vital signs, oxygen, CPR, splinting, and spinal immobilization.

4. Describe how PCR documentation changed in the 1980s and 1990s.

In the 1980s documentation reflected the established technician demonstrating the expanding clinical capabilities of the EMS provider, recording advanced airway procedures, medication administration, and cardiac arrest management. In the 1990s documentation reflected the established clinician. Patient Care Reports became more universal with many states mandating state-issued PCRs. Expanding clinical capabilities required more information. As the financing of EMS became more challenging, additional billing information was required.

5. Describe the impact of the White Paper on the development of EMS.

The White Paper identified traumatic injury as "the neglected disease of modern society," revealing the absence of medical care outside the walls of the nation's hospitals. The White Paper served as a wake-up call leading politicians, physicians, and policy makers to address the nation's inability to care for the sick and injured outside hospitals.

6. Describe the impact of the Emergency Medical Services Systems Act of 1973 on the development of EMS.

The Emergency Medical Services Systems Act of 1973 defined EMS systems and provided federal funding for the development of EMS systems.

7. Describe the relationship between PCR documentation and the advancement of the EMS profession through research.

Research validates the effectiveness of EMS practice and offers the opportunity to identify best practices. Today, most research occurs as a result of data collection and therefore the PCR is the primary research tool for EMS. In order to be valid, research requires accurate data, which the EMS provider is responsible to provide.

8. Describe the relationship between PCR documentation and the advancement of the EMS profession through quality management.

The quality of patient care is closely linked to the quality of documentation. The EMS provider's documentation should be an exact reflection of the care given to the patient. As was seen in the chapter case study, quality was questioned because the PCR failed to appropriately reflect the care given to the patient. Quality management efforts seek to improve quality of EMS care, and the PCR is the most valuable tool for evaluating the quality of patient care.

Chapter 3, page 31

1. Describe the link between documentation and reimbursement.

First, anytime EMS services are billed, an appropriately documented PCR must back up the billed services. Second, billing decisions are made based upon PCR documentation. Billing specialists are dependent upon the EMS professional to document the care given accurately so they can, in turn, make accurate billing decisions. Third, payment decisions are often made as a result of direct evaluation of PCR documentation. The PCR must appropriately reflect the patient care so an accurate payment decision can be made. Documentation and reimbursement are inseparable.

2. Describe the importance of the EMS professional having a basic understanding of Medicare history and process.

Having a basic understanding of Medicare's history promotes respect for this program, which is vital to EMS and the patients we serve. Having a basic understanding of Medicare process will enable the EMS professional to be more accurate in recording and documenting the patient's condition, assessment, and treatment.

3. Describe how medical necessity is established.

   Medical necessity is established from PCR documentation. Other factors such as dispatch mode and level of licensure of the responding ambulance determine the manner in which the EMS services are billed, but do not establish by themselves medical necessity.

4. Evaluate the PCR in Figure 3-2 and answer the following questions.

   * Based upon this PCR, how would you evaluate this documentation?
   * Based upon this PCR, how would you evaluate the patient care?
   * Based upon this PCR, would it be easy for billing staff to bill for the EMS services?
   * Place yourself at the desk of a Medicare Utilization Review nurse. Would you pay or deny the claim for these EMS services. Why?

   Evaluation: First, the PCR is fairly illegible and incomplete. The services provided and the reason the services were provided are not clear. Note the following:

   * *The patient's complete date of birth is missing.*
   * *The patient's past medical history and medications are not in agreement, suggesting the EMS professional failed to obtain the patient's past medical history.*
   * *Abbreviations are used in a manner that may provide a different interpretation of the patient's condition and treatment.*
   * *Interventions: Treatment times were not recorded; the medical reason for interventions and the patient's response to treatment are not clear.*
   * *The narrative fails to provide sufficient information regarding the patient's presentation, present illness, and treatment.*

   Based upon this documentation it would be easy to question whether the patient really received quality EMS care. Services presented in this PCR, as a claim billed to Medicare, would probably be denied: "Documentation does not support medical necessity."

5. Evaluate the PCR in Figure 3-3 and answer the following questions.

   * Based upon this PCR, how would you evaluate this documentation?
   * Based upon this PCR, how would you evaluate the patient care?
   * Based upon this PCR, would it be easy for billing staff to bill for the EMS services?
   * Place yourself at the desk of a Medicare Utilization Review nurse. Would you pay or deny the claim for these EMS services. Why?

   Evaluation: Note the differences between the PCR example in Figure 3-3 and the example in Figure 3-2. First, it is legible and clear. No data fields are left blank. Instead, "not applicable" supports that the EMS professional specifically addressed these areas in the history-taking process. All demographic and billing information is complete. Current medications and the patient's past medical history are in agreement. Abbreviations are used appropriately, leaving the reviewer with only one interpretation of the patient's condition and treatment. Treatment and the patient's response(s) to treatment are detailed and clear. Last, the narrative provides an appropriate summary of this EMS encounter. A reviewer, based upon the quality of this PCR documentation, would probably assume the patient received quality care. It would be easy for billing staff to code and bill for these EMS services appropriately. The EMS services presented in this PCR, as a claim billed to Medicare, would probably be paid.

6. Evaluate the PCR in Figure 3-4 and answer the following questions.

   * Based upon this PCR, how would you evaluate this documentation?
   * Based upon this PCR, how would you evaluate the patient care?
   * Based upon this PCR, would it be easy for billing staff to bill for the EMS services?
   * Place yourself at the desk of a Medicare Utilization Review nurse. Would you pay or deny the claim for these EMS services. Why?

   Evaluation: In this example, the advantages of an electronically generated PCR can be seen because there is no concern about legibility. However, a number of items are unclear:

- *Patient's chief complaint is "hip fracture." When did the fracture occur (acute vs. chronic condition)? Is the fracture stable or unstable?*
- *Destination facility is abbreviated. Abbreviations should never be used for origin and destination facilities. Instead, the names of the hospitals should be written out. Review staff may be located in another state and have difficulty in ascertaining whether the patient was transferred to the closest appropriate medical facility.*
- *Past medical history ("cardiac") is not in agreement with the patient's current medications.*
- *Treatment: Why did the patient receive oxygen, and what was the patient's response?*
- *The narrative fails to provide the reviewer with the essential information to summarize why the patient was transferred from the sending to the receiving facility.*

The EMS services presented in this PCR would be a challenge for billing staff to bill and, as a claim billed to Medicare, may be denied.

7. Evaluate the PCR in Figure 3-5 and answer the following questions.
   - Based upon this PCR, how would you evaluate this documentation?
   - Based upon this PCR, how would you evaluate the patient care?
   - Based upon this PCR, would it be easy for billing staff to bill for the EMS services?
   - Place yourself at the desk of a Medicare Utilization Review nurse. Would you pay or deny the claim for these EMS services. Why?

Evaluation: This example represents the manner in which EMS professionals document many interfacility transports. The tendency is to assume less information is required for a nonemergency. In truth, emergency and nonemergency EMS encounters require the same amount of documentation—all pertinent event and patient information must be recorded. The difference is not the amount of information obtained, but the type of information. A nonemergency transport, such as what is depicted in this example, illustrates the information that must be obtained for the nonemergency interfacility transport:

- *"Emergency Call" is listed in the Sending Facility field, representing a common mistake with the ePCR. This EMS professional pointed and clicked to a serious documentation error.*
- *Billing information and Medicaid number are blank, which could lead the coder to assume incorrectly that the patient had no coverage other than Medicare. Always verify the patient's health care coverage, and then record "not applicable" in these fields.*
- *"G Tube Replacement" is not an appropriate chief complaint. What was the patient's chief complaint? Did the patient have associated symptoms? Was the G tube replaced due to loss of patency? Did the patient remove the tube? "Unable to nourish patient due to loss of Gastric Feeding Tube" would reflect a more appropriate chief complaint if the patient were unable to verbalize pain or discomfort.*
- *Past medical history provides the clue as to why the patient has a G tube, but fails to provide the connection between past medical history and chief complaint.*
- *Current Medications and Allergies sections are blank, leaving the reviewer to assume the EMS provider failed to obtain the patient's medical history.*
- *No treatment was documented, leaving the reviewer to assume that transport by ambulance may not have been necessary.*
- *The narrative fails to provide any useful information regarding the medical necessity for the EMS services. More important, the narrative fails to summarize the patient's condition, assessment, treatment, and safety interventions.*

The EMS services presented in this PCR, as a claim billed to Medicare, would probably be denied as "documentation does not support medical necessity."

Chapter 4, page 45

1. List the four types of law and how they function within the legal system.
   - *Constitutional law, based upon the United States Constitution, establishes structure and function of the legal system.*
   - *Legislative law, established by the legislative branches of federal and state governments, produces laws and statutes.*

- *Common law seeks to apply previous court decisions to present legal cases.*
- *Administrative law establishes regulations set forth by federal and state agencies, adding detail to statutory law in the form of specific regulations that must be followed.*

2. List four things a plaintiff must prove in order to establish negligence.
   - *There was a duty to act according to a standard of care.*
   - *There was a failure to perform according to a standard of care.*
   - *The negligent act directly caused harm.*
   - *The negligent act also caused an actual loss.*

3. List and describe how standard of care is established.
   Standard of care, while usually based upon state law, is typically defined by written protocols, training curriculum, and standard operating procedures.

4. Why is accurate PCR documentation a standard of care?
   Documentation is a standard of care because:
   - *Accurate completion of a PCR is typically mandated by state law or local protocols.*
   - *Accurate completion of a PCR is what a reasonably prudent EMS professional would do as part of EMS practice.*

5. Describe the difference between scope of practice and standard of care.
   Scope of practice is usually established by state or county laws and defines the specific skills and parameters of practice for each level of EMS licensure. Standard of care is usually based on written protocols, training curriculum, and standard operating procedures. Standard of care is a ruler by which professional practice is measured against what a reasonably prudent peer would have done in the same situation.

6. You document "NKDA" (no known drug allergies) on a PCR, but you actually never asked the patient whether he or she was allergic to anything. Is this false documentation? Why or why not?
   Recording that a patient has no known allergies without specifically obtaining this information is false documentation. For that matter, documenting any medical information, whether a history, assessment, or examination finding, that was not specifically obtained is false documentation. If "NKDA" were simply noted in the medical record, the PCR must note that the information did not come directly from the patient. Example: "NKDA per patient medical record; unable to confirm due to patient's neurological status."

7. An EMT is found to have administered an IV medication that is prohibited in his or her state of practice. The EMT has violated:
   - Standard of care
   - *Scope of practice*

8. List the measures that can be used to guard a patient's protected health information in PCR documentation in order to comply with HIPAA regulations.
   - *Guard the PCR. If your organization uses a paper PCR, keep it secured at all times.*
   - *Know who is watching. If the PCR is left on a clipboard, it can be easily viewed by others. Keep the PCR out of the sight of anyone who is not allowed to view that patient's protected health information.*
   - *When providing a copy of the PCR to a medical facility, always give the copy directly to the person who will be assuming care of the patient. Never "drop it off" at the desk.*

9. You have an emergency patient and are being diverted to another facility. List the essential documentation elements related to EMTALA compliance.
   - *Document the hospital diversion thoroughly:*
     - *Diverting facility*
     - *Facility diverted to*
     - *Time of divert*
     - *Reason for divert*
     - *Person advising you of the divert*
     - *Medical control involvement (if applicable)*
     - *Supervisor notification*
   - *Inform the patient of the divert process and explain it to him or her. Document the patient's response using his or her own words, if applicable.*

- *If you questioned or verbalized disagreement over the change in receiving facility, document the details of this conversation.*
- *If the destination hospital is a greater distance away than the diverting facility, document the reason for not transporting the patient to the closest appropriate medical facility.*

10. Describe the purpose of the False Claims Act and how it relates to PCR documentation.
The False Claims Act is designed to protect the federal government from health care fraud by allowing claims to be made by individuals on behalf of the federal government.

Chapter 5, page 63

1. List the fundamentals of effective PCR documentation.
The effective PCR:
- *Represents the defining characteristics of the EMS professional*
- *Captures the EMS care given to the patient so EMS services can be billed accurately*
- *Reflects care that was within the boundaries of scope of practice and consistent with the standard of care*
- *Educates the non-EMS reviewer so the care and services are understood*
- *Integrates the EMS patient into the continuum of care at the receiving facility*

2. List three vital functions of PCR documentation.
The functions of EMS documentation are to inform, educate, and integrate the EMS patient into the care continuum.

3. List and describe the characteristics of clear documentation.
- *Clear documentation is legible.*
- *Clear documentation is not open to multiple interpretations, communicating only what is intended by the EMS provider who completed the PCR.*
- *Clear documentation appropriately uses abbreviations and avoids medical jargon.*

4. List the advantages, disadvantages, and principles associated with the use of medical abbreviations.
Advantages: The use of abbreviations represents a convenience in medical documentation and may speed up the documentation process.
Disadvantages: Abbreviations carry a risk of misinterpretation if the abbreviation used is not a "universal" abbreviation. Abbreviations also may not be understood by those in nonclinical roles, such as reimbursement.
Principles:
- *Utilize only agency-approved abbreviations.*
- *If abbreviations are used, an abbreviation key should be provided on the PCR.*
- *NEVER abbreviate the names of medication or units.*
- *NEVER use abbreviations on patient refusal forms.*
- *Avoid abbreviations in the diagnosis and Medical Necessity Statements.*
- *NEVER abbreviate the name of a medical facility.*

5. List and describe the characteristics of complete documentation.
- *Captures all pertinent aspects of the EMS encounter.*
- *Has an appropriate entry for all boxes, blanks, or data entry fields filled in. If a field is not pertinent to the EMS encounter, "not applicable" is written into the field.*
- *Answers the questions who, what, when, where, and how.*
- *Reflects balanced documentation. All pertinent data entry fields have entries, and the EMS encounter has been appropriately summarized in the narrative.*

6. List and describe the characteristics of correct documentation.
Correct documentation accurately represents a truthful and factual reporting of the EMS event as it happened. Correct documentation is free of spelling and grammatical errors and reflects proper management of any errors or omissions in documentation.

7. List the principles for correcting errors and omissions in PCR documentation.

Errors:

- *Do not attempt to "erase" or completely cross out a documentation error.*
- *Instead, draw one line through the error, enter the correction, and add your initials.*
- *Add an attachment or addendum to your PCR that notes your name and initials. In doing so, you are saying your initials equal your signature.*
- *Enter the date and time that you entered your correction.*
- *NEVER create another PCR after a completed PCR has been submitted.*

Omissions:

- *Use an addendum (supplement) to record the information that was omitted from the original PCR.*
- *Record the patient name, date, and incident number on the addendum or supplemental form.*
- *Write "Late Entry" and record the omitted information.*
- *Record the date and time of the late entry.*
- *Sign the entry.*

8. List and describe the characteristics of consistent documentation.
   Consistent documentation unifies the demographic, financial, clinical, and transfer of care components of the EMS encounter. Consistent documentation presents a picture of the EMS encounter in which event times, diagnosis, treatment, medical necessity, and the narrative are all in agreement.

9. List and describe the characteristics of concise documentation.
   Concise documentation is brief, recording only the essential facts relevant to the diagnosis, medical necessity, clinical care, and treatment.

Chapter 6, page 77

1. List the four data collection categories.
   - *D = Dispatch and Demographic Elements*
   - *A = Assessment of the EMS Event/Patient Elements*
   - *T = Treatment and Interventions Elements*
   - *A = Affirmation—Medical Necessity and Transfer of Care Statements*

2. List the essential documentation elements for reimbursement.
   - *Patient's demographic information*
   - *Primary method of payment*
   - *Insurance company information*
   - *Insured information*
   - *Worker's compensation information (if work related)*
   - *Name of closest relative information*
   - *Patient's (or policyholder's) employment information*

3. Describe the difference between diagnosis and chief complaint. What is the importance of this difference in documentation?
   A patient's chief complaint is what the patient tells you about his or her symptoms. The diagnosis is the conclusion made after evaluating the patient's chief complaint and performing a thorough assessment and examination. In documentation it is important to note that the patient's chief complaint and the diagnosis are not the same. The only exception is when the patient is unable to verbalize a chief complaint, in situations such as a cardiac arrest. To facilitate clearer documentation, "chief complaint" should be presented as "presenting problem."

4. Which of the following represents a chief complaint and which represents a diagnosis?
   - Multiple system trauma *(diagnosis)*
   - Fall *(diagnosis)*
   - G tube replacement *(neither)*
   - Hip fracture *(diagnosis)*
   - Headache *(chief complaint)*
   - Knee pain *(chief complaint)*

   For PCR documentation purposes, the chief complaints and diagnoses listed are all incomplete, and they lack appropriate description.
   - *Multiple system trauma—R/O Multiple System Trauma (Chest/Abdominal and Head Trauma, secondary to MVC)*

- *Fall—Neck pain, secondary to fall (standing position)*
- *G tube replacement—CVA (noncommunicative/total paralysis)—G tube replaced due to loss of patency*
- *Hip fracture—R/O Acute right hip fracture (extremity shortened with external rotation)*
- *Headache—Severe frontal headache with nausea (1 hour post onset)*
- *Knee pain—Acute nontraumatic knee pain*

5. List the critical documentation elements for IV therapy.
   - *Aseptic technique*
   - *Initiation time*
   - *Venipuncture site*
   - *Catheter*
   - *IV set*
   - *Number of attempts*
   - *Type of fluid*
   - *Infusion rate (infusion rate, total amount infused, and amount of fluid in bag at onset of transport (if preestablished)*
   - *Staff ID*
   - *Complications*
   - *Site evaluation at destination*
   - *Summarize intervention with a narrative comment*

6. List the critical documentation elements for medication administration.
   - *Medications administered prior to EMS arrival*
   - *Source of medical authorization*
   - *Allergy check*
   - *Administration time(s)*
   - *Name of medication(s)*
   - *Dose and units*
   - *Route of administration*
   - *Patient response*
   - *Staff ID*
   - *Summarize intervention with a narrative comment*

7. List the critical documentation elements for airway management.
   - *Airway procedure*
   - *Time of procedure*
   - *Equipment used*
   - *Airway grade*
   - *Number of attempts*
   - *Record whether procedure was successful*
   - *Response*
   - *Staff ID*
   - *Intubation adjuncts*
   - *Complications and rescue airway*
   - *ET confirmation (record, in detail, how placement of an advanced airway was confirmed)*
   - *Destination ET confirmation*
   - *Summarize intervention with a narrative comment*

8. List the critical documentation elements for patient safety.
   - *PPE used*
   - *Stretcher safety*

- *Patient belongings*
- *Summarize intervention with a narrative comment*

9. List the critical documentation elements for transfer of care.
   - *Receiving facility*
   - *Patient/incident disposition*
   - *Transport mode from scene*
   - *Condition of patient at destination*
   - *Destination type*
   - *Radio/phone report to destination facility*
   - *Transfer of care validation*

Chapter 7, page 105

1. List the functions of narrative documentation.
   - *Ties the EMS event together, allowing for highly descriptive information of the EMS event.*
   - *Affirms the medical necessity of the services provided.*
   - *Educates the reader as to why and how the EMS professional intervened. Remember, not everyone reading the PCR has an EMS background.*
   - *Explains exceptions.*
   - *Reinforces safety interventions.*
   - *Provides a place for documentation when specific fields are not provided in the PCR, such as a detailed physical examination.*

2. List the fundamental principles of narrative documentation.
   - *The narrative is written to provide descriptive detail of essential elements of the management of the EMS event.*
   - *The narrative must be focused, fulfilling a specific purpose and targeting areas of the EMS encounter.*
   - *The narrative must reflect the professionalism of the EMS provider.*
   - *The narrative must be written using appropriate medical terminology.*
   - *The narrative must educate. It must be written in an organized fashion that leads the person to an accurate conclusion regarding the EMS services provided.*
   - *The narrative must be signed, with signature, printed name, and title.*

3. List the essential elements of the Focused EMS Event Summary.
   - *EMS Diagnosis*
   - *Medical Necessity Statement*
   - *Scene Summary*
   - *Review of Systems*
   - *Intervention Summary*
   - *Changes/Response to Changes*
   - *Safety Summary*
   - *Disposition Summary*

4. List the essential documentation elements of the EMS Review of Systems.
   - *General/Appearance*
   - *Social History*
   - *Past Medical History*
   - *Mental Status*
   - *Neuro*
   - *Integumentary*
   - *HEENT*

- *Respiratory*
- *Cardiovascular*
- *Gastrointestinal*
- *Genitourinary*
- *Musculoskeletal*

5. Describe the purpose of the procedure note. Why is it important?

   Procedure notes capture essential information relating to a specific intervention, such as the initiation of an IV, spinal immobilization, or a safety precaution such as placing a patient in restraints. Procedure notes are important because they isolate information specific to the procedure (or intervention) and draw attention to the deliberate manner in which it was performed.

Chapter 8, page 119

1. Define medical necessity.

   Medical necessity is a reimbursement term that refers to whether a health care service is required (necessary) versus desired.

2. List the ground rules for this chapter's discussion of medical necessity.

   - *It is not the EMS professional's responsibility to establish medical necessity. The EMS professional's responsibility is to ascertain, through a thorough assessment and examination, the patient's history and condition and then accurately document so the EMS services can be evaluated for medical necessity.*
   - *Not every EMS event is medically necessary. Some transports are not medically necessary and should not be reimbursed. Therefore, in this discussion, we are neither attempting to learn to document in such a manner as to make EMS services medically necessary nor are we trying to make nonpayable EMS services payable by embellishing documentation in order to receive reimbursement.*
   - *PCR documentation must always be accurate. This discussion will not center on what to document but rather on the principles that guide the EMS professional to document medical necessity appropriately.*
   - *The examples given in this chapter are not to be used as documentation templates.*
   - *EMS services must never be denied or withheld based upon medical necessity.*

3. List and describe the Medicare coverage guidelines for EMS emergencies.

   Medicare considers an EMS encounter an "emergency" when a patient presents with an illness or injury of sudden onset that is characterized by severe symptoms for which the absence of immediate medical attention could place the patient's health in serious jeopardy. Medicare lists the following types of symptoms or conditions as that which could warrant EMS services:

   - *Severe pain or hemorrhage,*
   - *Unconsciousness or shock,*
   - *Injuries requiring immobilization of the patient,*
   - *Patient needs to be restrained to keep from hurting himself or others,*
   - *Patient requires oxygen or other skilled medical treatment during transportation, and*
   - *Suspicion that the patient is experiencing a stroke or myocardial infarction"*[1]

4. List and describe the Medicare coverage guidelines for nonemergencies.

   Medicare's coverage guidelines for nonemergency transportation are dependent upon:

   - *The patient's bed-confinement status.*
   - *The patient has a medical condition contraindicating transport by other means.*
   - *The patient has a condition that requires transportation by ambulance such as a patient who is combative or presents a danger to self or others.*
   - *Medicare guidelines stipulate that while bed-confinement is an important factor to determine the appropriateness of nonemergency ambulance transports by itself, it is by itself not the only determining factor that establishes medical necessity.*

[1] Department of Health and Human Services, Office of Inspector General, *Medicare Payments of Ambulance Transports* (January 2006), p. 3.

5. List the criteria that must be met for bed-confinement.

    To be considered bed-confined, the patient must be unable to get up from bed without assistance AND be unable to ambulate AND be unable to sit in a chair or wheelchair.

6. List the seven critical documentation fields for documentation of medical necessity.

    - *Demographics*
    - *EMS Diagnosis*
    - *Past Medical History*
    - *Assessment and Interventions*
    - *Focused EMS Event Summary*
    - *Signatures*
    - *Medical Necessity Statement*

7. List the purposes and advantages of the Medical Necessity Statement.

    The Medical Necessity Statement as a tool for EMS professional and billing staff serves the following purposes:

    - *Brings direct focus to the medical necessity for the EMS services in question*
    - *Provides structure for documenting medical necessity*
    - *Assists Medicare and other payers in making an informed payment decision*
    - *Provides for consistency in documentation of medical necessity*

8. List sources for obtaining a patient's past medical history.

    - *The patient (or family members)*
    - *The primary caregiver, usually the patient's primary RN at both the sending and receiving facilities*
    - *The medical records from sending and receiving facilities: the "H and P" (Health and Physical) and other medical history documentation from the sending facility*

9. List and describe the elements for evaluation and documentation of ambulation status.

    - *Does the patient walk independently?*
    - *What assistance is needed for the patient to ambulate (additional person, walker, or cane)?*
    - *Does the patient sit independently in a chair or wheelchair?*
    - *What is the patient's normal resting position?*
    - *Does the patient move independently?*
    - *How does the patient transfer from bed to stretcher?*
    - *How many assistants are used to move the patient, and how is the patient typically moved?*
    - *What type of assist devices does the patient use (walker, cane, or lift device)?*
    - *Does the patient use a wheelchair (be specific as to the type of wheelchair—conventional, bariatric, or Geri Chair)?*
    - *Is a safety device required to secure the patient in the wheelchair?*

10. List the questions that must be asked when evaluating medical necessity for interfacility transports.

    - *Does the patient meet Medicare's definition of "bed-confined"?*
    - *What is the patient's ambulation status?*
    - *What is the patient's transfer status?*
    - *What type of assist devices does the patient require?*
    - *What is the reason for medical treatment at the sending and receiving facilities?*

11. Describe the purpose of the Physician's Certification Statement (PCS).

    The Physician's Certification Statement (PCS) form provides physician's certification as to the medical necessity for nonemergency EMS service.

12. Describe the purpose of the Advance Beneficiary Notice (ABN).

    The Advance Beneficiary Notice (ABN) advises Medicare beneficiaries of the possibility that a nonemergency transport may not be covered by Medicare.

13. List and describe the EMS professional's ethical responsibilities in medical necessity documentation.
EMS professionals have profound ethical responsibilities in documentation of medical necessity:

- *Always document medical necessity accurately and truthfully, never documenting more or less than an accurate reflection of the patient's condition.*

- *Accurately document the EMS encounter and provide a summary statement specific to medical necessity. Remember, the EMS professional does not establish medical necessity but rather provides the critical information, in light of coverage requirements, so that medical necessity can be evaluated for billing and reimbursement purposes.*

- *EMS is called to respond at the discretion of the public. Callers request EMS services for a variety of reasons, including simply being in need of transportation to a hospital. Whether medical necessity is supported or not, the EMS professional must document medical necessity as it stands.*

- *Certain EMS services will not (and should not) be reimbursed. The EMS professional documents all EMS encounters accurately. EMS encounters that clearly fail to meet medical necessity requirements must be documented factually and accurately.*

Chapter 9, page 141

1. List and describe the EMS Documentation Process.

- *Apply the Five C's of Clinical Documentation—clear, complete, correct, consistent, and concise—to each Patient Care Report.*

- *Use the EMS DATA format—demographic, assessment, treatment, and affirmation data—to capture essential data for each EMS encounter.*

- *Assign the EMS Diagnosis.*

- *Summarize using the Focused EMS Event Summary.*

- *Complete the PCR with a Medical Necessity Statement.*

2. Discuss how time management skills impact documentation quality.
Time management is critical to consistent quality in PCR documentation. If PCR completion is habitually delayed to the end of the shift, the risk for errors and omissions is greatly increased. The opportunity for the highest quality in documentation is when the information is "fresh."

3. List the elements of the Focused EMS History.

- *Chief Complaint*
  - *Reason EMS Services Requested (state in patient's own words)*
- *History of Present Problem*
  - *Complete Description of Onset and Symptoms*
  - *Sequence of Events*
  - *Duration*
  - *Previous Events ("A Typical Event")*
  - *Changes in This Presentation*
- *Past Medical History (related to this illness)*
  - *General Health (patient's own words)*
  - *Medications*
  - *OTC Medications and Supplements*
  - *Related/Pertinent Hospitalizations/Surgeries*
  - *Major Illnesses*
- *Personal, Family, and Social History*
  - *Living Situation*
  - *Home/Economic Conditions*
  - *Occupation*
  - *Stressors*

4. List the elements of the Focused EMS Physical Exam.
   - *Vital Signs*
     - *BP/HR/RR/Oxygen Saturation and Temperature*
   - *General Appearance*
     - *Trauma/Appearance*
     - *Grooming*
     - *Emotional Status*
     - *Gestures/Body Language*
   - *Mental Status Evaluation*
     - *Speech (communication ability, quality/content)*
     - *Emotional State (depressed, angry, withdrawn)*
     - *Cognitive Abilities (memory, attention span)*
   - *HEENT*
     - *Trauma/Head (deformities)*
     - *Eyes (PERRLA)*
     - *Ears (drainage)*
     - *Nose (drainage)*
   - *Neurological Status*
     - *Trauma*
     - *EMS Cranial Nerve Exam*
     - *Motor Function/Sensory Function*
   - *Respiratory Status*
     - *Trauma/Appearance*
     - *Symmetry of Movement*
     - *Expansion/Excursion*
     - *Auscultation of Breath Sounds*
   - *Cardiovascular Status*
     - *Trauma*
     - *EKG/Rate/Rhythm*
     - *Heart Tones*
     - *Presence of JVD*
     - *Peripheral Edema/Peripheral Pulses*
   - *GI/GU Status*
     - *Trauma*
     - *Abdominal Appearance*
     - *Pain/Tenderness/Rebound/Referred Pain*
     - *Auscultation of Bowel Sounds*
     - *Change in Urinary Status*
   - *Musculoskeletal*
     - *Trauma/Deformities*
     - *Pain/Tenderness/Range of Motion*

Chapter 10, page 149

1. Describe how the EMS professional fulfills duty to act to the patient refusing EMS services.
   The EMS professional usually fulfills duty to act by simply treating and transporting the patient. However, when a patient refuses EMS services, fulfilling duty to act requires the EMS professional to evaluate the patient's capacity and competency while respecting his or her right to self-determination. If the EMS profes-

sional fails to ascertain that the patient is able to refuse, then duty to act is breached and the EMS professional can be accused of abandonment.

2. How does capacity differ from competency?
   Capacity refers to the patient's legal qualifications whereas competency refers to the patient's medical qualifications.

3. List the criteria that must be met for establishing competency.
   Competency is established when the patient meets all the following:
   - *Alert and oriented to person, place, time, and event*
   - *Has the ability to comprehend what is being communicated to him or her*
   - *Has the ability to understand the benefits of receiving EMS treatment*
   - *Has the ability to understand the risks of refusing EMS treatment and transport*
   - *Has the ability to make a decision based upon the information presented*

4. Describe how competency is determined in the refusal interview.
   The "A-B-C-D" approach for assessing competency is simple and applicable to EMS practice:
   - *A—Alert: Is the patient alert and oriented to person, place, time, and event?*

        AND

   - *B—Behavior: Is the patient's behavior appropriate?*

        AND

   - *C—Comprehension: Is the patient able to comprehend and understand his or her condition? Does the patient understand the illness, injury, treatment options, and risks associated with refusal of EMS service? Does the patient exhibit intact memory and judgment? Is the patient's speech normal? Is the patient cooperative? Is there any evidence of drug or alcohol use?*

        AND

   - *D—Decision: Is the patient legally qualified to make the refusal decision? Is the patient of legal age and able to make his or her own health care decisions? A patient that seems unwilling or unable to make a decision is showing an important diagnostic sign.*

5. Discuss why documenting "patient is alert and oriented" is insufficient.
   Documenting a patient as "alert and oriented" is insufficient because a competent patient will also demonstrate the ability to process information and make a decision.

6. List and describe the elements of a mental status exam.

   Examination:
   - *Ask the patient:*
     - *What is the season, year, and today's date?*
     - *Where are you right now?*
   - *Tell the patient: "I will name three objects." Example: "tree, car, house"*
     Ask the patient: "Please tell me the three objects I named for you."
   - *Ask the patient: "Spell world backwards."*
   - *Ask the patient: "Repeat the three objects that I named for you a minute ago."*
   - *Ask the patient to follow a command.*
     Example: "Place your right hand on top of your left hand."

   Evaluation:
   If the patient fails more than two out of the five evaluation steps, question competency.

7. List and describe the components of the EMS Refusal Interview.
   A—Assess Patient
   - *Chief complaint*
   - *Capacity*
   - *Competency*
   - *Consent for EMS care or refusal*
   - *Physical exam*

E—Educate Patient
- Condition
- Exam/assessment results
- Benefits of EMS care
- Education literature
- Limitations of EMS
- Answer questions

I—Inform Patient
- Risk and consequences

O—Offer EMS Services
- Offer three times—"3 Offer Rule"

U—Understanding
- Summarize refusal interview
- Validate patient understanding

8. List and describe the components of the CASE CLOSED Patient Refusal Narrative.

C = Document the patient's capacity and competency.

A = Document the results of the assessment and examination.

S = Document statements made by the patient reflecting the patient's competency.

E = Document that the patient was educated as to the patient's medical condition and the refusal process and how the patient validated understanding.

C = Document how the patient was instructed regarding the consequences of refusing EMS services and how the patient validated understanding.

L = Document instructing the patient of the limitations of EMS and the patient's verbalized understanding.

O = Document the number of times the patient was offered (and refused) EMS services.

S = Obtain appropriate signature(s) and initial(s).

E = Document any educational materials that were given to the patient—patient information sheets.

D = Document the patient and/or family was instructed to "dial 911 again" if the patient's condition changes or if the patient decides to accept EMS services.

Chapter 11, page 163

1. What type of events should be documented utilizing the Internal Communication Report?
Incident reporting is an essential tool for communicating events that could impact patient care, the EMS system, and the EMS staff. Examples include:
- Equipment issues
- Deviations from protocol or standard operating procedures
- Accidents and injuries
- Patient care errors
- Any incident that could result in a detrimental outcome for an EMS patient
- Any incident that could result in a detrimental outcome for the EMS system

2. List the Five Rights of Incident reporting.
- The Right Form
- The Right Facts
- The Right Focus
- The Right Forum
- The Right Follow-Up

3. What role does incident reporting have in risk management for EMS providers?
Incidents that could impact the EMS patient or EMS system require specialized documentation so that the associated legal risks can be appropriately managed. If incidents are not properly documented utilizing the right

form, communicating the right facts with the right focus, forum, and follow-up, EMS organizations face significant risk of litigation.

4. List and describe the general guidelines for incident reporting.
   General guidelines for the Internal Communication Report include:

   - *If the incident involved an equipment failure, document the name, model, and serial number of the equipment. Make sure to follow your organization's procedures for ensuring the device is not placed back in service. Document when and how the equipment was designated as being "out of service."*

   - *The ICR is for administrative use only. PCR documentation must be separate from incident reporting. Do not reference the presence of the ICR in PCR documentation.*

   - *The basic principles of documentation also apply to incident reporting. The report must be clear, complete, correct, consistent, and concise.*

   - *Complete the report as soon as possible after the event and promptly submit the form to your direct supervisor.*

   - *Include a descriptive narrative summarizing the incident and your immediate action in response to the issue.*

   - *If patient care was involved, reference the PCR and provide a description of any applicable patient responses, including statements made by the patient.*

   - *Make sure that all persons who witnessed the event are noted in the report (name, agency, and title).*

   - *Remember: The purpose of the report is to communicate an event so that proper investigation and resolution can take place.*

Chapter 12, page 173

1. List and describe the two legal concerns associated with verbal reporting.

   - *Patient confidentiality: The Health Insurance Portability and Accountability Act of 1996 (HIPAA) provides legal protection for the security and privacy of patient health information. Verbal reporting must always be conducted with respect to patients' protected health information. As much as possible, give and take Transfer of Care Reports in private to the nurse or physician who is assuming responsibility for the patient.*

   - *Inadequate care transfer: This presents a legal risk to the EMS professional. Failure to transfer responsibility for the patient's care appropriately to another qualified health care professional could be considered abandonment, and the EMS provider could be found negligent if there were a poor outcome.*

2. List and describe the three stages of EMS transfer of care.

   - *The Pre-Hospital Report: The Pre-Hospital Report, or radio/telephone report, alerts the receiving facility and introduces the patient to the next care providers. The Pre-Hospital Report provides only the basic information required to prepare for the patient's arrival.*

   - *The RN Transfer of Care Report: The RN Transfer of Care Report builds upon the Pre-Hospital Report and provides all necessary information for the receiving health care professional to assume responsibility and continue care.*

   - *The Patient Care Report (PCR): The PCR provides comprehensive written summarization of the entire EMS event as well as documented affirmation that the transfer of care occurred.*

3. Describe the purpose of the Pre-Hospital (radio) Report.
   The radio report serves an important function in EMS, tying pre-hospital care to hospital care, and represents the beginning of the transition out of EMS to in-hospital care. The radio report communicates the information to the receiving hospital that will best assist its staff in preparing for the patient.

4. *Describe the purpose of RN Transfer of Care Report.*
   The purpose of the Transfer of Care Report is to continue the process initiated by the Pre-Hospital Report, which introduced the patient to the receiving facility. The RN Transfer of Care Report serves two vital functions: it provides the details the nurse will need to continue care, and it introduces the patient to the next health professional who will be continuing care.

5. List and describe the purposes and characteristics of the Pre-Hospital Report, the Transfer of Care Report, and the PCR.

   - *Pre-Hospital (Radio/Phone) Report*
     - *Introduces the patient to the receiving facility for the purpose of preparation*

- *Brief*
- *Follows EMS DATA format*
- *Focused on information needed to prepare for care*
- *RN Transfer of Care Report*
  - *Introduces the patient to the next health care professional for the purpose of assuming legal responsibility and initiating care*
  - *Detailed, but focused*
  - *Focused on information needed to continue care*
  - *Communicates priorities*
  - *Follows DATA RN format*
- *Patient Care Report (PCR)*
  - *Documents all aspects of the EMS event*
  - *Comprehensive written report*
  - *Validates that proper transfer of care has occurred*

6. Describe the differences in reporting to the physician and the registered nurse.
The registered nurse is motivated by patient placement and flow through the hospital care system. The nurse requires information that will enable him or her to continue care and treatment, appropriately place the patient in the ED, arrange for placement as an inpatient if a direct admit, assign the appropriate acuity level for the patient, and anticipate the interventional needs of the patient. The physicians require information that will enable them to prioritize and prepare for interventions; facilitate physician specialists such as anesthesiology, surgery, or other specialized interventional services; prepare for invasive procedures; or prepare for emergent transport (such as air medical) if services are not available within the receiving medical facility.

7. List the elements of the EMS DATA Pre-Hospital Report format.
EMS = Service/Unit, Hospital and Incident Number, ETA
D = Patient Demographics
A = Chief Complaint and Assessment
T = Treatment
A = Acknowledgment

8. List the elements of the DATA RN format for Transfer of Care Reports.
D = Patient Demographics
  - *Name and age*
  - *Social history*
  - *Resuscitation status*
  - *Past medical history*
  - *History of present event*
A = Chief Complaint and Assessment Findings
T = EMS Treatment
A = Action Items/Priorities

R = Receiving MD/Sending MD
N = Nursing Priorities

Chapter 13, page 185

1. List the essential elements of crime scene documentation.
- *Description of the crime scene*
  - *Mechanism of injury*
  - *Patient presentation*
  - *Bystanders present at scene*
  - *Objects at scene*

- *Objects removed*
- *Clothing removed*
- *Items left at scene*
- *Statements made by the crime scene patient and bystanders*
- *Any report of unusual sounds*
- *Unusual scents or odors present*
- *Patient's social history*
- *Assessment/examination findings*
  - *Description of neurological status*
  - *Description of traumatic injury*

2. Describe the importance of documenting the neurological status of the crime scene patient.
The crime scene patient's mental status is critical information that must be meticulously documented. If the patient was neurologically intact upon EMS arrival and later deteriorates, the information provided and statements made by the patient will be crucial in the investigatory process.

3. List the essential documentation elements for the sexual assault patient.
- *The victim is part of the crime scene. Document the appearance of the patient and the surroundings prior to assessment or interventions.*
- *Document any clothing that was removed, why it was removed, and the disposition of the item.*
- *Document who was present with you during the course of evaluation, treatment, and transport.*
- *Document any item(s) that came into contact with the patient.*

4. List the essential documentation elements for the victim of child abuse.
- *Document who was caring for the child at the time EMS was called.*
- *Document who was present when the child was interviewed and examined.*
- *Document the onset time of the alleged illness/injury versus the time EMS was called.*
- *Document the child's general appearance and state of health.*
- *Document any sequential events leading up to the injury and the calling for EMS.*
- *Document exact conversations (with quotations) you had with the parent or caregiver.*
- *Describe the interaction between the child and parent or caregiver.*
- *Document the interview with the child, the questions asked, and how the child responded.*
- *Describe the child's injury, including location and characteristics of any markings (color, shape, and symmetry) on the child suggestive of injury.*
- *Document presence and characteristics of pain and the tool used for pain assessment.*

5. List the essential documentation elements for the victim of elder abuse.
- *Document a description of the patient's surroundings. Document exceptions such as unsanitary conditions, utilities not operational, inappropriate temperature for the season, the absence of prescribed medications, and inadequate food.*
- *Document who was with the patient upon EMS arrival.*
- *Document the general appearance of the patient. Does the patient seem well nourished and cared for?*
- *Document any odors, such as the smell of urine and/or feces.*
- *Document the presence of any trauma or injuries.*
- *Document the interaction between the patient and family members or caregivers.*

6. List the essential documentation elements for the patient with an order for physical restraints.
- *The order for transport hold and the physician's order for restraints must accompany PCR documentation, and the PCR should reference these attached documents.*
- *Document the specific manner in which the patient was restrained.*
- *Document neurovascular assessment per local practice guidelines.*
- *Document the manner in which the patient was educated regarding the transport hold and restraint procedures.*

- *Document the name of the physician ordering the hold or restraints and the time of the last physician and nursing assessments prior to departure.*

7. List the essential documentation elements for the suicidal patient.
   - *Document the patient's mood and affect. Describe the patient's surroundings and appearance in detail.*
   - *Document the presence of any friends, family members, or bystanders at the scene.*
   - *Document the presence of any suicide note and document the disposition of the note.*
   - *Document the mechanism used. If the patient or mechanism was moved, document the original location and where the object was relocated. In cases when the patient has been successful in the attempt, if pertinent objects relating to the attempt (medication bottles, weapons or rope, etc.) were not moved or disturbed, make it clear in documentation that the scene was left undisturbed.*

8. List and describe the three types of advance directive.
   - *Living will: The living will specifies the treatment the patient desires to refuse (or accept) in the event of being unable to make the decision him- or herself.*
   - *Health care proxy: The health care proxy designates a person to make treatment decisions for the patient in the event of being incapacitated.*
   - *Durable power of attorney for health care decisions: The durable power of attorney for health care decisions also designates a person to make treatment decisions but extends authority to the designated individual to make final treatment decisions, including cessation of treatment.*

Answers to
Critical Thinking Discussion Exercises

Chapter 1, page 1

1. How does PCR documentation represent EMS as a profession?

   Consider: The EMS Patient Care Report is to EMS as:

   - *The physician's evaluation and management documentation is to the medical profession.*
   - *The nursing care plan and nursing diagnosis is to the nursing profession.*
   - *The documents of judicial process are to the legal profession.*

   Most (nonemergency) physicians, nurses, attorneys, and other professionals will never ride along on an "ambulance run." These professionals and those in the reimbursement professions lack a proper understanding of the EMS profession. What they know of EMS will be gleaned mostly from PCR documentation. As EMS providers become more proficient in communicating the uniqueness of the profession through the Patient Care Report, EMS will become more defined as a profession.

2. The average paramedic spends 1,000 to 1,300 hours training. Of these 1,000 to 1,300 hours, it could be estimated that about 5 to 10 hours are spent in documentation skill training. Describe how this impacts documentation skill development.

   Consider: In order to become proficient in documentation, the EMS professional at every level of licensure must receive adequate instruction and training in documentation. The lack of training in documentation reveals EMS has yet to embrace documentation as an EMS skill. Hence, the industry continues to struggle with issues directly tied to PCR documentation.

Chapter 2, page 13

1. Describe your current attitude toward quality management. How do you generally respond to negative feedback on your patient care, and how does this relate to proficiency in documentation?

   Consider: It has been easy in the past to view quality management as the "quality police," out to get you at the least minute documentation infraction. This represents an unhealthy attitude for the EMS professional. Even when quality management performs overzealously, isn't it better for someone within the organization to review your patient care than a plaintiff's attorney or the review board at the state EMS office? If this is your attitude toward quality management, attaining proficiency in documentation is part of the answer. Most experienced EMS providers are not bothered by someone evaluating their immobilization or assessment skills. Time and effort were invested in attaining proficiency. Confidence is a by-product of efficiency.

2. What is the difference between a profession and a vocation? Does EMS qualify as a profession? Are EMS skills professional skills or technical ("trade") skills?

   Consider: A vocation is often considered to be a nonprofessional occupation when it lacks the ability to self-govern and does not possess specialized knowledge requiring formal education for entry into the field. Technical (trade) skills are usually associated with a vocation, whereas professional skills are expressions of the attributes of a profession. PCR documentation records the unique professional activities of the EMS provider. Through documentation, the EMS professional transforms clinical judgments and interventions into a professional, legal, and financial document.

Chapter 3, page 31

1. Reimbursement decisions are often made by medical review nurses. Is this a problem for EMS, and what impact does this have on PCR documentation?
   Consider: The medical review nurse usually has very limited understanding of EMS and may not comprehend the different levels of licensure, scope of practice, and the capabilities of EMS. This fact of reimbursement explains why payment decisions often baffle EMS administrators and underscores the pressure that is on the PCR to communicate EMS services appropriately.

2. In thinking through the process of a PCR being evaluated for medical necessity, does documentation of patient care ever need to educate?
   Consider: Having established the lack of understanding of EMS that often exists in the reimbursement process, documentation must not only inform but also educate. This is especially important in documenting medical necessity.

3. What value would there be in EMS professionals spending time in a billing department? Likewise, would there be value in billing staff riding with field crews or would this lead to a conflict of interest?
   Consider: The central problem in "billing versus clinical" is the lack of understanding that "clinical" has of "billing," and vice versa. This is unfortunate because the mission of the entire EMS organization is the same: delivery of great patient care. The value in EMS professionals spending time with billing staff, and billing staff riding with crews is the building of relationships. Successful EMS organizations are built upon strong relationships. Building relationships between departments increases understanding of what is needed to fulfill the mission.

Chapter 4, page 45

1. "Negligence in documentation is negligence in patient care." Do you agree with this statement? Why or why not?
   Consider: Negligence is neglecting to perform professional duties in the same manner as another reasonably prudent EMS provider. Because accurate documentation of patient care is a standard of care, negligence in documentation equals negligence in patient care.

2. What are the ramifications when scope of practice and standard of care are ill-defined? How does this affect documentation?
   Consider: Well-defined scope of practice and standard of care provide EMS personnel with clear boundaries for professional practice. Well-defined practice boundaries make it seasier for the EMS professional to provide and document patient care.

Chapter 5, page 63

1. Discuss how the Five C's of Clinical Documentation impact:
   - Professionalism in documentation
   - Financing of EMS
   - Legal accountability in documentation

Consider:

1. *Professionalism in documentation*—A clear, complete, correct, consistent, and concise PCR displays the attributes of the EMS profession.
   - *National Identity:* Each PCR represents potential for extractable data for research through data collection. Data collection is a small step in the development of a national identity.
   - *Specialized Knowledge and Education:* Each PCR represents potential to display the EMS professional's specialized knowledge and education.
   - *Autonomy:* Each PCR provides the EMS professional the opportunity to display the differences in EMS practice from medicine, nursing, and other public safety professions.
   - *High Ethical Standards:* Each PCR must represent the integrity of the EMS professional. Accurate and truthful PCR documentation is a by-product of high ethical standards, which provide restraint from embellishing documentation, adjusting times, and simply trying to look good on paper.

2. *Financing of EMS—A clear, complete, correct, consistent, and concise PCR assists those involved in the reimbursement process in making accurate billing and payment decisions.*
3. *Legal accountability in documentation—A clear, complete, correct, consistent, and concise PCR validates that patient care met the standard of care and was delivered within the boundaries of scope of practice and in compliance with state and federal laws.*

Chapter 6, page 77

1. What is the difference between documenting "head-to-toe exam—no findings" and documenting positive or negative findings for each anatomical area?
   Consider: Documenting a positive or negative finding for each anatomical area demonstrates that the EMS professional performed a thorough examination. "No findings" is a dangerous documentation practice because it can be suggestive of an incomplete assessment. It is quite easy for an assessment to reveal "no findings" when you didn't bother to look.

2. Why is it important to document care performed by other health care providers prior to your arrival, such as spinal immobilization, airway management, IV therapy, and medication administration?
   Consider: Failing to document care rendered by another health care provider transfers responsibility to the EMS provider receiving the patient. Not only do you receive the patient, but you also assume the legal risks associated with improper care, unless documented in the PCR. Here are a few examples:

   - *Spinal Immobilization: Was the patient ambulating prior to immobilization and what position was the patient in when immobilized? Was the cervical collar size chosen appropriate, and what were the results of the patient's motor and sensory examination pre/post immobilization?*
   - *Airway Management: Did the patient require suctioning? Did you observe appropriate BVM technique on your arrival? Was the oral airway size appropriate for the patient?*
   - *IV Therapy: Is the IV patent and is the site secure? What is the appearance of the IV site?*
   - *Medication Administration: Were allergies checked prior to administration? If the patient is receiving a medication infusion, was the rate verified? Is the infusion compatible with other infusions?*

Chapter 7, page 105

1. Evaluate each of the common narrative formats: SOAPIER, CHARTE, Head-to-Toe, and Review of Systems. What are the strengths and weaknesses of each of these formats? What information is captured by the Focused EMS Event Summary that the other formats leave out?
   Consider: Make a list of the information that is captured by each narrative format, and compare each to the Focused EMS Event Summary.

2. Why is the use of a consistent narrative format important?
   Consider: One of the challenges in review of PCR documentation for quality, administrative, or reimbursement purposes is the inconsistencies from one EMS provider to another. Thinking globally, consider a Medicare carrier's medical review staff who reviews thousands of EMS encounters from all the EMS organizations in its jurisdiction and the challenges associated with deciphering each EMS provider's unique approach to documentation.

3. The statement "EMS professionals do not diagnose" has been accepted since the birth of EMS. Is this true for today's EMS practice? Is it harmful to the advancement of the EMS profession to discount the diagnostic capabilities of the EMS professional?
   Consider: Failing to acknowledge the diagnostic capabilities of the EMS professional reinforces the perception of EMS as only a technical vocation. Therefore, EMS skills can be mechanically performed apart from critical thinking. Holding on to this paradigm retains a mind-set that prohibits the EMS professional from advancing into a professional role. Each time you perform an assessment and critically think through your findings to arrive at a treatment plan, you've made a diagnosis somewhere in the process. Interventions are usually based upon a diagnosis. Johnny and Roy (from the television show Emergency!) assessed, reported their findings, and then followed orders. Today, EMS professionals assess, diagnose, and make treatment decisions. As an EMS professional, you diagnose. Don't let anyone talk you out of it.

4. What is the difference between the EMS Diagnosis and the medical diagnosis?

   Consider: The EMS Diagnosis is the diagnosis made by the EMS provider, based upon the EMS assessment in the pre-hospital environment. The medical diagnosis is the physician's diagnosis that ultimately defines the patient's condition and directs medical treatment. Often the EMS Diagnosis and the medical diagnosis are the same, but other times they will differ.

Chapter 8, page 119

1. Does a negative attitude toward nonemergencies or interfacility transports have any effect on how medical necessity is documented?

   Consider: The EMS professional's role in medical necessity documentation is not to establish medical necessity but rather to obtain and document the essential elements of medical necessity. If the EMS professional embraces an attitude such as "I only transport patients that really need an ambulance," it will be easy not to obtain the information or to document with prejudice.

2. What are the dangers of EMS services being paid incorrectly? Which is worse: the EMS organization not getting paid or the EMS organization being paid incorrectly?

   Consider: The goal is not reimbursement, but obtaining the proper reimbursement. If a medically necessary transport was inappropriately denied, time will be spent in appealing the payment decision. If an EMS organization exhibits patterns of questionable billing practices, it will be at risk for federal scrutiny.

3. When a claim for EMS services is denied, "documentation does not support medical necessity," is Medicare stating the EMS services were not necessary?

   Consider: No. Although a medical necessity determination can be a subjective decision by a medical review nurse lacking understanding of EMS, this common denial reason is stating only that the documentation failed to make the case for medical necessity.

Chapter 9, page 141

1. What is the relationship between proficiency in assessment skills and proficiency in documentation?

   Consider: You cannot give what you do not possess. The EMS professional will never document beyond his or her current skill level.

2. Apart from documentation, list the primary clinical skills or procedures applicable to your level of licensure. How could sharpening each of these skills positively impact your documentation?

   Consider: The deeper the clinician's knowledge base and skill level, the greater the opportunity for descriptive documentation. Sharpening clinical skills provides the opportunity to communicate your expertise through documentation.

Chapter 10, page 149

1. Does your EMS organization have a culture that encourages obtaining patient refusals? How does this affect the quality of patient care and documentation?

   Consider: Patient refusals represent the area of highest legal risk in EMS, and many EMS organizations fail to manage the risk. If documentation practices within an organization consistently fail to communicate appropriate management of patient refusals, it is likely a culture exists that encourages refusals. Eventually, there will be a poor outcome. Will you be the one that takes the fall for a widespread problem?

2. Does the attitude of the EMS professional toward a patient presenting with only the appearances of a minor illness or injury impact patient care and documentation?

   Consider: The EMS professional must approach each patient from a position of equality, regardless of the presentation. Otherwise, it is possible for EMS professionals to "triage" patients by their presenting chief complaint or injury, prior to assessment. If you ever believe a patient is wasting your time, you're on dangerous ground.

Chapter 11, page 163

1. A patient was dropped while being transferred from the hospital bed to the ambulance stretcher. Evaluate the following statement on the EMS provider's Incident Communication Report: "The patient fell while

being lifted from the hospital bed onto the ambulance stretcher. The patient fell because the nurse failed to lock the wheels on the hospital bed."

Consider: First, we hope that considerable additional detail was provided in the Incident Communication Report, such as positioning of the bed and stretcher, number of staff moving the patient, technique used to move the patient, and assessment of the patient after the injury. This statement, however, is subjective, placing blame on another care provider instead of simply documenting the facts of the incident. Did the nurse really fail to lock the wheels, or did the locking mechanism on the hospital bed fail? Document the facts of the incident and leave conclusions for the investigation.

Chapter 12, page 173

1. What are the advantages of mandated report formats for verbal reporting?
   Consider: Universal formats have advantages for both EMS and hospital personnel. First, universal formats provide EMS with a uniform structure for reporting. The EMS provider will know what information is required in reporting. Second, universal formats train hospital staff to listen for key information. Third, universal formats provide consistency among EMS providers in a jurisdiction.

2. Why is it helpful for the EMS professional to identify nursing priorities in the Transfer of Care Report?
   Consider: Identifying nursing priorities assists the RN in establishing priorities so an immediate plan of care can be initiated for the patient. It also builds respect between the EMS and the nursing professional.

Chapter 13, page 185

1. Why is the patient's social history relevant to PCR documentation of the crime scene patient?
   Consider: The patient's social history often provides clues into the events leading to the incident in question. Capturing this information in the PCR not only is essential to continuity of care but also can be beneficial to law enforcement's investigation of the incident.

2. What types of patient encounters might an EMS professional have bias toward? Is there any correlation between personal bias and the manner in which certain EMS patient encounters are documented (for example, the patient with a behavioral emergency)?
   Consider: Unfortunately, none of us is immune from bias. The EMS professional must become aware of any personal biases toward specific patient groups and take steps to become professionally neutral.

Glossary

abandonment: The termination of the EMS professional–provider relationship prior to an appropriate transfer of care.

acuity: The complexity of a patient's condition.

Advance Beneficiary Notice (ABN): A CMS-mandated form advising patients (and family members) that Medicare may deny claims for services that may not be deemed reasonable and necessary.

advance directive: A written statement directing a person's health care treatment in advance.

autonomy: The professional ability to work independently from others.

Balanced Budget Act 1997 (BBA): Legislation that mandated payment for EMS services according to a "fee schedule" and drastically changed the funding of EMS.

capacity: Legal qualification to make a health care decision.

competency: Ability, in the medical sense, to make a health care decision.

concise: Brief but complete.

condition codes: Codes that assign a number designation to a diagnosis and are used by EMS providers to communicate the patient's condition in broad categories.

continuity of care: Seamless transitions in the delivery of health care from one provider, or department, to another.

credible: Believable and trustworthy.

data elements: Categories of information obtained for each EMS event that provide documentation of clinical care and EMS performance.

dataset: Groups of data that describe every aspect of an EMS event.

Diagnosis Related Groups (DRGs): A method of predetermined payment based on a patient's diagnosis.

durable power of attorney for health care decisions: A type of advance directive that grants full legal authority for treatment decisions to a designated person if the patient is incapacitated.

duty to act: The EMS professional's legal responsibility to treat a patient in accordance with EMS regulations and standards of care.

electronic Patient Care Report (ePCR): A computer-based PCR that provides for enhanced data collection, quality management, and system integration.

Emergency Medical Treatment and Active Labor Act of 1986 (EMTALA): Known as the "antidumping" law, EMTALA legislation seeks to prevent hospitals from turning away patients, either by the denial of care and treatment or through inappropriate transfers to other facilities.

EMS Diagnosis: The medical conclusion (diagnosis) made by the EMS professional as a result of assessment and examination.

False Claims Act: A federal law designed to discourage fraud in federal health care programs by allowing an individual to file suit on behalf of the government against those suspected of fraudulent practice.

Federal Health Care Program: A government-funded program that provides health insurance benefits. Examples include Medicare and Medicaid.

fee for service reimbursement: A method of reimbursement that pays the health care provider for each specific service or intervention that is provided to the patient.

Focused EMS Event Summary: A documentation format that summarizes the EMS event by focusing documentation on critical elements of the EMS event.

health care proxy: A type of advance directive that designates a person to make treatment decisions for the patient in the event of being incapacitated.

Health Insurance Portability and Accountability Act of 1996 (HIPAA): HIPAA legislation is designed to protect the privacy and security of health care information.

ICD9: The International Classification of Diseases (ICD) system classifies diseases into specific categories and assigns a number—the ICD9 code—to each classified disease. ICD9 codes are commonly referred to as diagnosis codes.

implied consent: Consent for medical treatment that is assumed for a patient who is unable, by reason of illness or injury, to grant informed consent.

incident report: Report of any event that is inconsistent with established guidelines, procedures, or outcomes.

informed consent: Consent for medical treatment that is given after the patient has been fully educated.

involuntary consent: Consent for medical treatment that is made on behalf of a patient by legal process.

living will: A type of advance directive that stipulates the types of treatment that the patient desires to refuse (or accept) in the event of being unable to make the decision him- or herself.

malpractice: Negligence committed by a professional.

medical necessity: A reimbursement term that defines whether or not a health care service is required versus desired.

Medicare Compliance Programs: Formal programs adopted by health care organizations that provide quality measures, education, and administrative oversight to ensure their practices reflect government standards.

Medicare fee schedule: Provides standardized payment rates for EMS services based upon seven levels of service.

narrative: Open text field(s) that allows for recording of information deemed important by the EMS provider.

negligence: The failure, resulting in harm or injury, to perform an expected duty.

patient advocate: An advocate is someone who acts on behalf of another.

Patient Care Report (PCR): The professional documentation tool of the EMS professional. Previously known as the "trip sheet," the PCR records demographic, financial, and patient care information. Also referred to as the "Pre-hospital Care Report."

Physician's Certification Statement (PCS): The PCS form is a CMS-mandated form that requires physicians to certify medical necessity for the nonemergency patient.

procedure codes: Also known as CPT/HCPCS codes. Assign a numerical designation to a medical procedure. In EMS, procedure codes are used to define the level of service, such as BLS, ALS, and Specialty Care Transport (SCT).

procedure notes: Notes that capture essential information relating to interventions.

Prospective Payment System (PPS): A health care payment system that predetermines payment for a health care service.

reimbursement: The manner in which EMS systems and providers are paid.

risk management: A function of leadership within an organization that seeks proactively to identify, manage, and reduce the risks associated with everyday EMS activities.

scope of practice: Defined boundaries for professional practice.

self-determination: A person's right to make his or her own decisions without pressure or inappropriate influence.

standard of care: Detailed written guidelines or standards that direct patient care.

subsidies: Financial sources other than Medicare, Medicaid, or insurance companies; usually from city or county taxes.

tertiary care facility: A health care facility that is able to provide the highest level of specialized diagnostic and treatment service.

time management: A developed skill that enables a person to get more value out of his or her time.

tort: A type of civil law that seeks to provide monetary relief for harm or damages, which are usually a result of negligence.

Index